W9-AMK-819

Matrices for Statistics

Matrices for Statistics

M. J. R. Healy

Professor of Medical Statistics
University of London

CLARENDON PRESS · OXFORD

Oxford University Press, Walton Street, Oxford OX2 6DP
Oxford New York Toronto
Delhi Bombay Calcutta Madras Karachi
Petaling Jaya Singapore Hong Kong Tokyo
Nairobi Dar es Salaam Cape Town
Melbourne Auckland

and associated companies in
Berlin Ibadan

Oxford is a trade mark of Oxford University Press

Published in the United States
by Oxford University Press, New York

© M. J. R. Healy, 1986

First published 1986
First published in paperback (with corrections) 1991

All rights reserved. No part of this publication may be reproduced, stored in a retrieval system, or transmitted, in any form or by any means, electronic, mechanical, photocopying, recording, or otherwise, without the prior permission of Oxford University Press.

This book is sold subject to the condition that it shall not, by way of trade or otherwise, be lent, re-sold, hired out, or otherwise circulated without the publisher's prior consent in any form of binding or cover other than that in which it is published and without a similar condition including this condition being imposed on the subsequent purchaser.

British Library Cataloguing in Publication Data
Healy, M. J. R.
Matrices for statistics.
1. Mathematical statistics 2. Matrices
I. Title
519.5'01'5129434 QA276
ISBN 0-19-852207-X
ISBN 0-19-852248-7 (Pbk)

Library of Congress Cataloging in Publication Data
Healy, M. J. R.
Matrices for statistics.
Bibliography: p.
Includes index.
1. Matrices. 2. Mathematical statistics. I. Title.
QA188.H43 1986 512.9'434 86-753
ISBN 0-19-852207-X
ISBN 0-19-852248-7 (Pbk)

Printed in Great Britain by Dotesios Ltd., Trowbridge, Wilts

Preface

The role of mathematics in statistics is not dissimilar to that of statistics in science as a whole. A large number of scientists and technologists use statistics in their research, for the design of their studies and for the analysis of their findings, and their use will be more effective if they know something of the reasoning that underlies the textbook methods that they employ. They will have no pretensions to be statisticians themselves and will consult professional statisticians if difficult or nonstandard problems of design or analysis occur in the course of their work. In the same kind of way, the majority of statisticians will mainly be concerned with the applications of pre-existing statistical methods and will only from time to time have to undertake the mathematical task of devising new techniques. They will, however, need enough mathematics to understand the techniques which they use and also to read and absorb the specialized statistical literature in which new methodology is published.

The result of this is that statisticians and users of statistical methods require something rather different from the standard mathematical textbook. With any given type of mathematics, such as linear algebra, group theory, or numerical analysis, they may need to become acquainted with quite advanced topics; but they will need less of the detail at more elementary levels. Moreover, they will need to be fairly selective in their study of a topic which will at best be secondary to their main interests.

This book has been written with these considerations in mind. It has been kept as brief as I have found reasonably consistent with the coverage which I felt was useful. I am well aware that whole books have been written on topics which are covered here in a single chapter; the enthusiasts, and the statisticians who are mainly concerned with the development of theory, can find there the extra material which they will need in their work. My hope is that what is included here will help the users of statistical methods to obtain a deeper insight into the mathematical underpinning of the techniques which they rely on.

London
1985

M. J. R. H.

Contents

1
Introducing matrices

1.1 Introduction

In many branches of applied mathematics, and especially in statistics, we need to deal with numerical quantities laid out in a rectangular array of rows and columns, otherwise known as a *two-way table* or *matrix.* Some examples are shown in Fig. 1.1. Figure 1.1(a) shows a (rather small) *data matrix*, with measurements of height, weight, and arm circumference for each of 10 subjects. Each of the 10 rows of the matrix corresponds to one subject and contains his three measurements; each of the three columns

Height (cm)	Weight (kg)	Arm circ (cm)
167.9	71.8	30.0
183.8	75.1	29.4
172.9	58.0	26.0
175.5	58.4	25.7
176.4	67.7	27.9
168.5	75.2	31.7
178.0	67.3	27.4
178.0	71.3	29.0
175.4	75.9	30.3
171.2	65.3	29.4

(a) Data matrix

$$\begin{bmatrix} 208.74 & 27.14 & -24.59 \\ 27.14 & 386.42 & 100.32 \\ -24.59 & 100.32 & 32.94 \end{bmatrix}$$

(b) Symmetric matrix of sums of squares and products

$$[176.76 \quad 68.60 \quad 28.68]$$

(c) Row-vector of means

Fig. 1.1 Matrices and vectors

$$\begin{bmatrix} 3 & 0 & 0 & 0 \\ 0 & 5 & 0 & 0 \\ 0 & 0 & -1 & 0 \\ 0 & 0 & 0 & 6 \end{bmatrix}$$

$$= [3 \quad 5 \quad -1 \quad 6]^{\mathrm{d}}$$

Fig. 1.2 Diagonal matrix

corresponds to one of the measurements and contains its values for the ten subjects. Figure 1.1(b) contains the sums of squares and products about the means derived from the data in Fig. 1.1(a). This is a *square matrix* with the same number of rows as columns, and it is *symmetric* about its *main diagonal*, the one running from top left to bottom right. The matrix in Fig. 1.1(c) has only one row, and is known as a *row vector*. The numbers in a matrix are known as its *elements*.

It is often convenient to think of the elements of a vector as the geometrical coordinates of a point and of the vector itself as corresponding to this point or to the line drawn to it starting at the origin. This geometric analogy helps to clarify many of the algebraic results, especially for those who are prepared to imagine spaces with more than three dimensions, and we shall use it frequently.

The purpose of matrix manipulations is to enable us to handle arrays of this kind as single entities without having to specify in repetitive detail what happens to the individual elements. To do this we shall generally use capital letters to denote matrices and lower-case for ordinary numbers. We need a name for these to distinguish them from matrices; we shall refer to them as *scalars*.

1.2 Some notation and definitions

Consider a general matrix denoted by A. We shall write a_{ij} for the element in the ith row and jth column of A so that the whole matrix can be written in the form

$$\begin{bmatrix} a_{11} & a_{12} & \dots & a_{1n} \\ a_{21} & a_{22} & \dots & a_{2n} \\ \dots & \dots & \dots & \dots \\ a_{m1} & a_{m2} & \dots & a_{mn} \end{bmatrix}.$$

m×n

This matrix has m rows and n columns, and we call it an $m \times n$ matrix. It is occasionally useful to denote the same matrix by $[a_{ij}]$, putting the typical element in brackets. The data matrix in Fig. 1.1(a) is a 10×3 matrix; the sums-of-squares-and-products matrix in Fig. 1.1(b) is a 3×3 square matrix; the vector in Fig. 1.1(c) is actually a 1×3 matrix.

A vector is a matrix with only one row (a *row vector*) or only one column (a *column vector*). It turns out to be convenient to assume that all vectors used are column vectors unless otherwise stated. We shall often need to distinguish vectors from general matrices, and we shall do so by denoting them by bold lower-case letters. Thus $\boldsymbol{c}$ would stand for the column vector

$$\boldsymbol{c} = \begin{bmatrix} c_1 \\ c_2 \\ \vdots \\ c_m \end{bmatrix}.$$

In fact, $\boldsymbol{c}$ is an $m \times 1$ matrix. Note that the matrix A can be thought of as being made up from m row vectors or from n column vectors.

We have already met square matrices with $m = n$, and among these symmetric matrices for which $a_{ij} = a_{ji}$. A more special case still is that of a square matrix whose elements are all zero except for those on the main diagonal (Fig. 1.2). This is called a *diagonal matrix*. We can take the elements and put them into a vector, $\boldsymbol{a}$ say, and we shall then denote the diagonal matrix by $\boldsymbol{a}^{\mathrm{d}}$. (This notation, due to K. D. Tocher, deserves more widespread use.)

If all the diagonal elements of a diagonal matrix are equal, the matrix is called a *scalar matrix*; we shall see that scalar matrices behave very like scalar (nonmatrix) quantities. In particular, a scalar matrix all of whose diagonal elements are equal to 1 is a *unit matrix* and is denoted by I (we may have to write I_p if we need to emphasize that it is a $p \times p$ matrix). Any matrix all of whose elements are equal to zero is a *zero matrix* and will be

$$I_2 = \begin{bmatrix} 1 & 0 \\ 0 & 1 \end{bmatrix} \quad I_4 = \begin{bmatrix} 1 & 0 & 0 & 0 \\ 0 & 1 & 0 & 0 \\ 0 & 0 & 1 & 0 \\ 0 & 0 & 0 & 1 \end{bmatrix} \quad \mathbf{1}_3 = \begin{bmatrix} 1 \\ 1 \\ 1 \end{bmatrix}$$

$$0_{3 \times 2} = \begin{bmatrix} 0 & 0 \\ 0 & 0 \\ 0 & 0 \end{bmatrix} \quad 0_{2 \times 5} = \begin{bmatrix} 0 & 0 & 0 & 0 & 0 \\ 0 & 0 & 0 & 0 & 0 \end{bmatrix} \quad \mathbf{0}_4 = \begin{bmatrix} 0 \\ 0 \\ 0 \\ 0 \end{bmatrix}$$

Fig. 1.3 Unit and zero matrices and vectors

square main diagonal

$$\begin{bmatrix} 4 & 0 & 0 & 0 \\ 7 & 5 & 0 & 0 \\ 3 & 2 & 6 & 0 \\ 1 & 4 & 8 & 7 \end{bmatrix} \qquad \begin{bmatrix} 3 & 2 & 6 \\ 0 & 5 & 7 \\ 0 & 0 & -8 \end{bmatrix}$$

Fig. 1.4 Triangular matrices

denoted by 0, with suffices for the numbers of rows and columns where necessary.

We often need to refer to vectors all of whose elements are 1's or 0's. We shall denote these by **1** and **0** respectively, assuming as usual that they are column vectors unless we state otherwise. Examples of unit and zero matrices and vectors are shown in Fig. 1.3.

A *triangular matrix* is (slightly confusingly) a square matrix with all the elements above, or below, the main diagonal equal to 0; we speak of a lower or upper triangular matrix, respectively. Two such matrices are shown in Fig. 1.4. It should be carefully noted that, for brevity, a symmetric matrix is sometimes written with the elements either above or below the main diagonal omitted. Do not confuse this with a triangular matrix.

It may be worth pointing out that a 1×1 matrix is square, symmetric, triangular, diagonal, and scalar. For all practical purposes it is interchangeable with the scalar quantity consisting of its single element (computer programmers may compare this with the difference between a real quantity and an array of length 1).

1.3 Simple operations on matrices

It is important to realize that we are entitled to *define* operations such as addition and multiplication of matrices in any way that seems good to us. What we have to do is to find operations that we are likely to find useful and to take advantage of any analogies between these and the more familiar operations on scalars or ordinary arithmetic. Many of them turn out to be entirely obvious. We should start by defining what we mean by the *equality* of two matrices. We do this by writing $A = B$ if and only if A and B are both of the same size and $a_{ij} = b_{ij}$ for all i and j. It is worth noting, though, that if A and B are both (say) 6×8 matrices, the simple statement $A = B$ is a shorthand version of 48 separate equality statements between scalar quantities.

We also define *addition* of matrices in the obvious way. The matrix $A + B$ is only defined when A and B are the same size, and then $C = A + B$ with $c_{ij} = a_{ij} + b_{ij}$ for all i and j. *Subtraction* of matrices is defined in exactly the same way. Examples are shown in Fig. 1.5.

An operation which has no scalar counterpart is *transposition*, which is the interchange of rows and columns. If A is a $m \times n$ matrix, then we call B

$$\begin{bmatrix} 1 & 7 & 4 \\ 3 & 2 & -1 \\ 5 & 4 & 8 \end{bmatrix} + \begin{bmatrix} 2 & 9 & 7 \\ 2 & 2 & 6 \\ 1 & 3 & -4 \end{bmatrix} = \begin{bmatrix} 3 & 16 & 11 \\ 5 & 4 & 5 \\ 6 & 7 & 4 \end{bmatrix}$$

$$\begin{bmatrix} 2 & 8 & 6 \\ 3 & 4 & 4 \\ 7 & 5 & 9 \end{bmatrix} - \begin{bmatrix} 3 & 8 & 7 \\ 3 & 2 & 4 \\ 4 & 5 & 5 \end{bmatrix} = \begin{bmatrix} -1 & 0 & -1 \\ 0 & 2 & 0 \\ 3 & 0 & 4 \end{bmatrix}$$

$$\begin{bmatrix} 8 & 6 \\ 2 & 7 \\ 5 & 4 \end{bmatrix}^{\mathrm{T}} = \begin{bmatrix} 8 & 2 & 5 \\ 6 & 7 & 4 \end{bmatrix}$$

$$4\begin{bmatrix} 2 & 1 & 8 \\ 3 & -4 & 7 \end{bmatrix} = \begin{bmatrix} 8 & 4 & 32 \\ 12 & -16 & 28 \end{bmatrix}$$

Fig. 1.5 Simple matrix arithmetic

the *transpose* of A if B is an $n \times m$ matrix with $b_{ij} = a_{ji}$ for all i and j. We shall write A^{T} for the transpose of A (the notation A' is sometimes used). A 3×2 matrix and its transpose are shown in Fig. 1.5. Note that a symmetric matrix can be defined as one which is equal to its own transpose. Because we normally assume that $\boldsymbol{a}$ denotes a column vector, we shall indicate where necessary that a vector is a row by writing it as $\boldsymbol{a}^{\mathrm{T}}$.

We shall come to multiplication of matrices in the next section, but we can define here what we mean by multiplying a matrix by a scalar. If A is a matrix and k a scalar quantity, we define the product kA to be the matrix B, the same size as A, with $b_{ij} = ka_{ij}$ for all i and j. An example is shown in Fig. 1.5. Note that $A + (-1 \cdot B) = A - B$ and $A + A = 2A$ just as we would wish.

1.4 Matrix multiplication

Very many of the matrix applications in this book will be to do with the solution of simultaneous equations, in particular of the so-called normal equations that arise in the solution of multiple regression problems by least squares. Consider the following m simultaneous equations in n unknowns $x_1, x_2, \ldots, x_n$:

$$\begin{aligned} a_{11}x_1 + a_{12}x_2 + \cdots + a_{1n}x_n &= h_1, \\ a_{21}x_1 + a_{22}x_2 + \cdots + a_{2n}x_n &= h_2, \\ &\cdots \\ a_{m1}x_1 + a_{m2}x_2 + \cdots + a_{mn}x_n &= h_m, \end{aligned}$$

and notice that we can write them as the equality of two column vectors, $\boldsymbol{l} = \boldsymbol{h}$, where $\boldsymbol{h}^{\mathrm{T}} = [h_1, h_2, \ldots, h_m]$ and the ith element of $\boldsymbol{l}$ is the sum

$$a_{i1}x_1 + a_{i2}x_2 + \cdots + a_{in}x_n$$

(a scalar quantity, of course). The commas separating the h_i have the same significance as spaces. Now the left-hand side of the equations is a combination of the coefficients a_{ij} and the unknowns x_j. The coefficients arrange themselves very naturally into a matrix $A = [a_{ij}]$ with m rows and n columns, and we can set out the x's, like the h's, in a column vector $\boldsymbol{x}$ with n elements. We now define the product $A\boldsymbol{x}$ to be the left-hand side vector $\boldsymbol{l}$. Notice that this product is only defined when $\boldsymbol{x}$ has as many elements as A has columns, and that when an $m \times n$ matrix multiplied by an $n \times 1$ vector produces an $m \times 1$ vector, with the middle dimension, the one common to the two matrices, 'cancelling out'. Now we can write the equations in the very compact form

$$A\boldsymbol{x} = \boldsymbol{h}.$$

It is not difficult to extend this definition. Suppose that X is a matrix with n rows, an $n \times p$ matrix say. We can think of it as being made up of p column vectors each with n elements. Multiplying each of these vectors by A produces p product vectors each with m elements, and we can put these together to make an $m \times p$ matrix, H say, and write $AX = H$.

Here we have our full definition of matrix multiplication. The product AB is defined only if B has the same number of rows as A has columns. If then A is $m \times n$ and B is $n \times p$, the product AB is equal to C where C is an $m \times p$ matrix with

$$\begin{aligned} c_{ij} &= a_{i1}b_{1j} + a_{i2}b_{2j} + \cdots + a_{in}b_{nj} \\ &= \Sigma_k a_{ik}b_{kj}. \end{aligned}$$

This is the sum of products of the ith row of A by the jth column of B. Notice again the 'cancelling' of the middle suffix.

This rather peculiar multiplication of matrices behaves in some ways like the multiplication of scalars. For example, it is *distributive* with addition, in that (fairly obviously) $(P + Q)R = PR + QR$, that is, assuming that all the matrices concerned are of suitable sizes so that the different operations are properly defined; we shall leave the reader to make this proviso in future. It is also not difficult to verify that multiplication of matrices is *associative* in that $(AB)C = A(BC)$ so that we can write the triple product ABC without ambiguity. Moreover we can link together the two definitions of multiplication that we have used by pointing out that, if k is a scalar and K a scalar matrix with all its diagonal elements equal to k, then $kA = KA$.

One familiar feature of scalar multiplication, however, does not carry over to the matrix case. For the product AB to be defined, it is necessary

$$\begin{bmatrix} 3 & 2 & 7 \\ 4 & 1 & 6 \end{bmatrix} \begin{bmatrix} 1 & 4 \\ 3 & 2 \\ 5 & 8 \end{bmatrix} = \begin{bmatrix} 44 & 72 \\ 37 & 66 \end{bmatrix}$$

$$\begin{bmatrix} 4 & 7 \\ 8 & 3 \end{bmatrix} \begin{bmatrix} 5 & 2 \\ 4 & 6 \end{bmatrix} = \begin{bmatrix} 48 & 50 \\ 52 & 34 \end{bmatrix}$$

But

$$\begin{bmatrix} 5 & 2 \\ 4 & 6 \end{bmatrix} \begin{bmatrix} 4 & 7 \\ 8 & 3 \end{bmatrix} = \begin{bmatrix} 36 & 41 \\ 64 & 46 \end{bmatrix}$$

Fig. 1.6 Matrix multiplication

that B should have as many rows as A has columns; the fact that A comes first and B second matters, because it may well be that the product BA is not defined at all. Even if it is—if say, A is $m \times n$ and B is $n \times m$—AB will be $m \times m$ and BA will be $n \times n$ so that in general they will be different sizes. Finally, even if $m = n$ so that A and B are square and AB and BA are the same size, a little experimentation will show that their corresponding elements may not be equal to each other—see Fig. 1.6 for an example. Matrix multiplication is, in general, *noncommutative*. This means that we need to be a bit careful when we expand matrix expressions. For example

$$(A + B)(A + B) = AA + AB + BA + BB,$$
$$(A + B)(A - B) = AA + BA - AB - BB.$$

We can write A^2 and B^2 instead of AA and BB but we cannot in general simplify the right-hand sides any further.

The effects of multiplying on the left and right by a diagonal matrix are worth noting. Suppose that A is $p \times p$ and K is $p \times p$ and diagonal. Then AK is $p \times p$ and its columns are those of A each multiplied by the corresponding diagonal element of K; KA is also $p \times p$, but now its rows are multiples of the rows of A. Figure 1.7 shows an example.

$$\begin{bmatrix} 3 & 2 & 7 \\ 4 & 5 & 1 \\ 6 & 3 & 3 \end{bmatrix} \begin{bmatrix} 1 & 0 & 0 \\ 0 & 3 & 0 \\ 0 & 0 & 2 \end{bmatrix} = \begin{bmatrix} 3 & 6 & 14 \\ 4 & 15 & 2 \\ 6 & 9 & 6 \end{bmatrix}$$

$$\begin{bmatrix} 1 & 0 & 0 \\ 0 & 3 & 0 \\ 0 & 0 & 2 \end{bmatrix} \begin{bmatrix} 3 & 2 & 7 \\ 4 & 5 & 1 \\ 6 & 3 & 3 \end{bmatrix} = \begin{bmatrix} 3 & 2 & 7 \\ 12 & 15 & 3 \\ 12 & 6 & 6 \end{bmatrix}$$

Fig. 1.7 Multiplication by a diagonal matrix

Some matrices do in fact commute. In particular, if A is square and K is scalar with all its diagonal elements equal to k, then $KA = kA = AK$. As special cases, if I is a unit matrix and 0 a square zero matrix, then $IA = AI = A$ and $0A = A0 = 0$, with the unit and zero matrices behaving like unit and zero scalars respectively. These relations actually hold when A is rectangular of size $m \times n$ rather than square, but to be precise we should write $K_m A = AK_n$ to show that scalar matrices of different sizes are needed for multiplication on the right and left.

What is the transpose of a matrix product, $(AB)^{\mathrm{T}}$ say? If $AB = C$ we know that $c_{ij} = \Sigma_k a_{ik} b_{kj}$, and if $(AB)^{\mathrm{T}} = D$ then $d_{ij} = c_{ji} = \Sigma_k a_{jk} b_{ki}$, the sum of products of the jth row of A by the ith column of B. But this is the same as the sum of products of the ith row of B^{T} by the jth column of A^{T}. It follows that

$$(AB)^{\mathrm{T}} = B^{\mathrm{T}} A^{\mathrm{T}}.$$

Note that when one product exists, then so does the other.

1.5 Partitioned matrices

It is often convenient to consider a matrix as being made up of smaller matrices—we shall call these *submatrices*. If A is 6×8, for example, we might wish to write it in the partitioned form

$$A = \begin{bmatrix} A_{11} & A_{12} \\ A_{21} & A_{22} \end{bmatrix}$$

where A_{11} and A_{12} might have five rows with A_{21} and A_{22} the remaining three, while A_{11} and A_{21} have, say, two columns and A_{12} and A_{22} the remaining six. We can even write $A = [A_{ij}]$ if the sizes of all the submatrices are obvious from the context.

For some purposes we can treat the submatrices rather as if they were scalar elements. If B is the same size as A and partitioned in exactly the same way, it is easy to see that

$$A + B = [A_{ij} + B_{ij}].$$

Less obviously it can be shown that, provided all the products are defined,

$$AB = \begin{bmatrix} A_{11}B_{11} + A_{12}B_{21} & A_{11}B_{12} + A_{12}B_{22} \\ A_{21}B_{11} + A_{22}B_{21} & A_{21}B_{12} + A_{22}B_{22} \end{bmatrix}$$

using the usual row-times-column rule for matrix multiplication. If desirable, we can of course partition the rows or the columns of the matrix into three or more parts, and the same properties will continue to hold.

1.6 Sums of squares and products

If $\boldsymbol{x}$ is a column vector of size $m \times 1$, then $\boldsymbol{x}^{\mathsf{T}}\boldsymbol{x}$ is a 1×1 matrix which is interchangeable for most purposes with the scalar quantity which is its single element. If the elements of $\boldsymbol{x}$ are $x_1, x_2, \ldots, x_m$, then $\boldsymbol{x}^{\mathsf{T}}\boldsymbol{x} = \Sigma_i x_i^2$, the sum of the squares of the elements. In our geometrical interpretation, the elements of $\boldsymbol{x}$ are the coordinates of a point and then $\boldsymbol{x}^{\mathsf{T}}\boldsymbol{x}$, by Pythagoras' theorem, is the squared distance of the point from the origin.

In exactly the same way, if $\boldsymbol{y}$ is another $m \times 1$ vector, $\boldsymbol{x}^{\mathsf{T}}\boldsymbol{y} = \boldsymbol{y}^{\mathsf{T}}\boldsymbol{x} = \Sigma_i x_i y_i$, the sum of products of the elements of the two vectors. This too has a geometrical interpretation. We find that $\boldsymbol{x}^{\mathsf{T}}\boldsymbol{y}/(\boldsymbol{x}^{\mathsf{T}}\boldsymbol{x} \cdot \boldsymbol{y}^{\mathsf{T}}\boldsymbol{y})^{\frac{1}{2}} = \cos\theta$ where θ is the angle between the lines joining the origin to the two points represented by $\boldsymbol{x}$ and $\boldsymbol{y}$.

Sums of squares and products are basic to many statistical techniques. Suppose X is a data matrix, with a row for each item (specimen, subject, individual, . . .) and a column for each variate. Then the whole matrix of (crude) sums of squares and products is simply $X^{\mathsf{T}}X$. We can generalize this at once to weighted sums of squares and products—if $\boldsymbol{w}$ is a vector of weights with an element for each item and $W = \boldsymbol{w}^{\mathsf{d}}$ is a diagonal matrix with the elements of $\boldsymbol{w}$ on its diagonal, then the weighted sums of squares and products are the elements of $X^{\mathsf{T}}WX$.

1.7 Linear combinations and contrasts

If $x_1, x_2, \ldots, x_n$ denote n variates, we often wish to consider a set of m linear combinations of them:

$$\begin{aligned} y_1 &= c_{11}x_1 + c_{12}x_2 + \ldots + c_{1n}x_n, \\ y_2 &= c_{21}x_1 + c_{22}x_2 + \ldots + c_{2n}x_n, \\ &\ldots \\ y_m &= c_{m1}x_2 + c_{m1}x_2 + \ldots + c_{mn}x_n. \end{aligned}$$

In matrix terms, we can describe the vector $\boldsymbol{y}$ as a *linear transformation* of the vector $\boldsymbol{x}$ and write $\boldsymbol{y} = C\boldsymbol{x}$ with $C = [c_{ij}]$. From elementary statistical theory we know that the means of the y_i's are given by

$$\mathsf{E}y_i = c_{i1}\mathsf{E}x_1 + c_{i2}\mathsf{E}x_2 + \cdots + c_{in}\mathsf{E}x_n$$

and if $\operatorname{cov}(x_i, x_j) = v_{ij}$ then

$$\operatorname{cov}(y_i, y_j) = \Sigma_k \Sigma_l c_{ik} c_{jl} v_{kl}.$$

This can all be put into matrix notation. The coefficients c_{ij} have already been assembled into a matrix C, and we can put the expected values into vectors written as $\mathsf{E}\boldsymbol{x}$, $\mathsf{E}\boldsymbol{y}$ and the variances and covariances into matrices V_{xx} and V_{yy} (these will be symmetric matrices of size $n \times n$ and $m \times m$

respectively). Then we have the important relationships

$$\mathsf{E}y = C\,\mathsf{E}x, \qquad V_{yy} = CV_{xx}C^{\mathsf{T}}.$$

We can look at the product $y = Cx$ in a slightly different way. Above we described y as a vector of linear combinations of the x's with coefficients in the matrix C. From another point of view, the column vector y is a linear combination of the columns of C with the x's as coefficients. In just the same way, the row vector $z^{\mathsf{T}} = x^{\mathsf{T}}C$ is a linear combination of the rows of C.

The linear combination $y_j = \Sigma_i c_{ij} x_i$ is called a *contrast* if the sum of the coefficients, $\Sigma_i c_{ij}$, is zero. We can write this condition as $c_j^{\mathsf{T}}\mathbf{1} = 0$ where c_j is the vector of coefficients and $\mathbf{1}$ is a vector each of whose elements is equal to 1. If the x's are uncorrelated variates with variance σ^2, then the variance of y_j is $c_j^{\mathsf{T}}c_j\sigma^2$. If $y_k = c_k^{\mathsf{T}}x$ is another contrast between the same x variates, then the variance of y_k is $c_k^{\mathsf{T}}c_k\sigma^2$ and the covariance of y_j and y_k is $c_j^{\mathsf{T}}c_k\sigma^2$. The correlation between y_j and y_k is thus $c_j^{\mathsf{T}}c_k/(c_j^{\mathsf{T}}c_j \cdot c_k^{\mathsf{T}}c_k)^{\frac{1}{2}}$ and this is the cosine of the angle between the two vectors c_j and c_k. If $c_j^{\mathsf{T}}c_k = 0$, the contrasts are uncorrelated and the vectors, in the geometrical interpretation, are at right angles to each other. The two contrasts and their vectors are then said to be *orthogonal*. If the vectors are of unit length, i.e. if $c_j^{\mathsf{T}}c_j = 1 = c_k^{\mathsf{T}}c_k$ we say that they are *orthonormal*.

1.8 Orthogonal transformations

Consider a $p \times p$ square matrix A which is such that every pair of columns consists of two orthonormal vectors (such a thing exists—indeed, we show two in Fig. 1.8). We call this an *orthogonal* matrix. Its defining property can be neatly expressed in matrix terms as $A^{\mathsf{T}}A = I$.

Let x_1 and x_2 be two vectors and consider the linear transformations $y_1 = Ax_1$ and $y_2 = Ax_2$ with A an orthogonal matrix. Then

$$\begin{aligned} y_1^{\mathsf{T}}y_1 &= x_1^{\mathsf{T}}A^{\mathsf{T}} \cdot Ax_1 \\ &= x_1^{\mathsf{T}} \cdot I \cdot x_1 \\ &= x_1^{\mathsf{T}}x_1. \end{aligned}$$

In the same way we find that $y_2^{\mathsf{T}}y_2 = x_2^{\mathsf{T}}x_2$ and $y_1^{\mathsf{T}}y_2 = x_1^{\mathsf{T}}x_2$. We can express this by saying that the lengths of the vectors and the angle between them are *invariant* under *orthogonal transformation*. In the geometric interpretation, an orthogonal transformation corresponds to a rotation of the coordinate axes which leaves the relative positions of the origin and the two points corresponding to the vectors unchanged.

$$\begin{bmatrix} 0.5000 & 0.5000 & 0.5000 & 0.5000 \\ -0.6708 & -0.2236 & 0.2236 & 0.6708 \\ 0.5000 & -0.5000 & -0.5000 & 0.5000 \\ 0.2236 & -0.6708 & 0.6708 & -0.2236 \end{bmatrix}$$

$$\begin{bmatrix} 1/\sqrt{4} & 0 & 0 & 0 \\ 0 & 1/\sqrt{2} & 0 & 0 \\ 0 & 0 & 1/\sqrt{6} & 0 \\ 0 & 0 & 0 & 1/\sqrt{12} \end{bmatrix} \begin{bmatrix} 1 & 1 & 1 & 1 \\ -1 & 1 & 0 & 0 \\ -1 & -1 & 2 & 0 \\ -1 & -1 & -1 & 3 \end{bmatrix}$$

Fig. 1.8 Orthogonal matrices

1.9 The trace of a matrix

If A is square, the *trace* of A (the German term is *spur*) is simply the sum of the diagonal elements—we denote it by $\operatorname{tr} A$. When all the relevant products exist we have $\operatorname{tr}(PQ) = \operatorname{tr}(QP)$ and $\operatorname{tr}(PQR) = \operatorname{tr}(QRP) = \operatorname{tr}(RPQ)$.

1.10 Direct products

We have defined one kind of matrix multiplication, but other operations known as 'products' are sometimes found to be useful. One of these is the so-called *direct* or *Kronecker* product, usually denoted by the symbol $\otimes$.

Suppose A is $m \times n$ and B is $p \times q$. Then the direct product $A \otimes B$ is of size $mp \times nq$ and is most easily described as the partitioned matrix

$$\begin{bmatrix} a_{11}B & a_{12}B & \dots & a_{1n}B \\ a_{21}B & a_{22}B & \dots & a_{2n}B \\ \dots & \dots & \dots & \dots \\ a_{m1}B & a_{m2}B & \dots & a_{mn}B \end{bmatrix}.$$

This matrix product is useful in describing certain kinds of factorial design. Suppose we have two factors A and B, with two and three levels respectively, and that we summarize their effects by sets of orthogonal contrasts. We can set out the coefficients of the contrasts as the rows of two matrices:

possible choices would be

$$A = \begin{bmatrix} 1 & 1 \\ -1 & 1 \end{bmatrix} \quad \text{and} \quad B = \begin{bmatrix} 1 & 1 & 1 \\ -1 & 0 & 1 \\ 1 & -2 & 1 \end{bmatrix}.$$

$$\text{Then } A \otimes B = \begin{bmatrix} 1 & 1 & 1 & 1 & 1 & 1 \\ -1 & 0 & 1 & -1 & 0 & 1 \\ 1 & -2 & 1 & 1 & -2 & 1 \\ -1 & -1 & -1 & 1 & 1 & 1 \\ 1 & 0 & -1 & -1 & 0 & 1 \\ -1 & 2 & -1 & 1 & -2 & 1 \end{bmatrix}.$$

The rows of this matrix (after the first) give the coefficients of the contrasts defining the main effects and interaction in the two-factor experiment involving A and B. A particular application of this occurs in biological assay. Here the levels of A are unknown versus standard preparation, and the levels of B are three equally spaced log doses. Then the contrasts defined by the rows of the product matrix after the first are those usually labelled slope, curvature, materials, parallelism, and opposing curvature respectively. Note that all these contrasts are mutually orthogonal.

Examples 1

1.1 Suppose X $(n \times p)$ is a data matrix with n cases and p variates. Let $\mathbf{1}$ be a vector consisting of n 1's. Show that

(a) The column sums of X are in the vector $\mathbf{1}^{\mathrm{T}}X$.
(b) The matrix of deviations from column means is

$$X - \mathbf{1}\mathbf{1}^{\mathrm{T}}X/\mathbf{1}^{\mathrm{T}}\mathbf{1}.$$

(c) The matrix of sums of squares and products of deviations from column means is

$$(X - \mathbf{1}\mathbf{1}^{\mathrm{T}}X/\mathbf{1}^{\mathrm{T}}\mathbf{1})^{\mathrm{T}}(X - \mathbf{1}\mathbf{1}^{\mathrm{T}}X/\mathbf{1}^{\mathrm{T}}\mathbf{1})$$

and this is equal to

$$X^{\mathrm{T}}X - X^{\mathrm{T}}\mathbf{1}\mathbf{1}^{\mathrm{T}}X/\mathbf{1}^{\mathrm{T}}\mathbf{1},$$

the crude sums of squares and products minus the corrections for the means.

(d) To transform X to deviations from column means, we multiplied in (b) by the matrix $(I - \mathbf{1}\mathbf{1}^{\mathrm{T}}/\mathbf{1}^{\mathrm{T}}\mathbf{1})$. What happens if we multiply again by the same matrix? In other words, what is $(I - \mathbf{1}\mathbf{1}^{\mathrm{T}}/\mathbf{1}^{\mathrm{T}}\mathbf{1})^2$? [You should be able to obtain your answer first and only then check it by working out the square of the matrix explicitly.]

1.2 If A is the symmetric matrix

$$\begin{bmatrix} 4 & 8 & 4 \\ 8 & 25 & 11 \\ 4 & 11 & 30 \end{bmatrix}$$

and U is the upper triangular matrix

$$\begin{bmatrix} 2 & 4 & 2 \\ 0 & 3 & 1 \\ 0 & 0 & 5 \end{bmatrix}$$

verify that U is a kind of square root of A, in that $A = U^{\mathrm{T}}U$.

1.3 If $A = \begin{bmatrix} 1 & 3 & 5 \\ 2 & 4 & 7 \\ 8 & 9 & 3 \end{bmatrix}$,

and

$$P = \begin{bmatrix} 0 & 1 & 0 \\ 1 & 0 & 0 \\ 0 & 0 & 1 \end{bmatrix}, \qquad Q = \begin{bmatrix} 1 & 0 & 0 \\ 0 & 3 & 0 \\ 0 & 0 & 1 \end{bmatrix}, \qquad R = \begin{bmatrix} 1 & 2 & 0 \\ 0 & 1 & 0 \\ 0 & 0 & 1 \end{bmatrix}$$

calculate PA, QA, and RA. What are the effects of these multiplications on the rows of A? By what operations could P, Q, and R be derived from unit matrices?

1.4 If $A = \begin{bmatrix} 0 & 0 & 6 \\ \frac{1}{2} & 0 & 0 \\ 0 & \frac{1}{3} & 0 \end{bmatrix}$,

calculate A^2, A^3, and $(A^2)^2$.

1.5 If $P = \begin{bmatrix} 6900 & 3240 & 840 & 9420 \\ 3240 & 6480 & 9072 & 1944 \\ 840 & 9072 & 15288 & -3192 \\ 9420 & 1944 & -3192 & 14100 \end{bmatrix} / 21384$

and $X = \begin{bmatrix} 10 & 785 \\ 8 & 162 \\ 7 & -266 \\ 12 & -607 \end{bmatrix}$,

form PX.

1.6 Let $P = \begin{bmatrix} 3 & 4 \\ 1 & 2 \end{bmatrix}$, $Q = \begin{bmatrix} 2 & 3 \\ 5 & 6 \end{bmatrix}$, $R = \begin{bmatrix} 4 & 1 \\ 2 & 2 \end{bmatrix}$.

Form PQ, PR, and QR and verify that

$$\text{(a)} \ (P + Q)R = PR + QR, \qquad \text{(b)} \ (PQ)R = P(QR).$$

Show that, although $PQR \neq QRP$, the traces of the two matrices are equal.

1.7 If A is $m \times n$ and B is $n \times p$, what are the vectors $B\mathbf{1}$ and $(AB)\,\mathbf{1}$? Show how to check the calculation of the product AB by comparing $(AB)\,\mathbf{1}$ with $A(B\mathbf{1})$.

1.8 Let A be $p \times q$, let $\boldsymbol{x}$ be $q \times 1$, and let $\boldsymbol{y} = A\boldsymbol{x}$. Each element of $\boldsymbol{y}$, say y_i, is a function of the elements of $\boldsymbol{x}$. Show that $\partial y_i/\partial x_j = a_{ij}$. We can write $\partial \boldsymbol{y}/\partial \boldsymbol{x} = A$.

In the same way, let B be symmetric $q \times q$ and $y = \boldsymbol{x}^{\mathrm{T}}B\boldsymbol{x}$. Then $\partial y/\partial x_j = 2(b_{1j}x_1 + b_{2j}x_2 + \cdots + b_{pj}x_p) = 2\boldsymbol{b}_j^{\mathrm{T}}\boldsymbol{x}$ where $\boldsymbol{b}_j$ is the jth column of B. We can write $\partial y/\partial x = 2\boldsymbol{x}^{\mathrm{T}}B$.

1.9 Suppose that the vector $\boldsymbol{y}$ contains the values of a dependent variate y which is to be predicted by a set of x-variables whose values are in the columns of a matrix X. If the regression coefficients are in a vector $\boldsymbol{b}$ the residuals are the elements of the vector $\boldsymbol{y} - X\boldsymbol{b}$ and the residual sum of squares is given by

$$(\boldsymbol{y} - X\boldsymbol{b})^{\mathrm{T}}(\boldsymbol{y} - X\boldsymbol{b}).$$

To minimize this, we must differentiate with respect to the b's and equate the results to zero. Show that this leads to the *normal equations*

$$(X^{\mathrm{T}}X)\boldsymbol{b} = X^{\mathrm{T}}\boldsymbol{y}.$$

2
Determinants

2.1 Introduction

After defining addition, subtraction, and multiplication of matrices, it would be natural to go on to define division, but first we must make a substantial detour. In scalar division, we always have to be careful to avoid dividing by zero, an operation which is not defined. It turns out that the similar problem with matrix division is more complicated; it is not enough to avoid dividing by zero matrices, we must beware of matrices which have a zero 'value' in a rather general sense. Of course, the question 'What is the value of a matrix?' has no answer; a matrix is no more than a shorthand notation for its elements and does not have any one number as its value. Nevertheless, a particular number can be associated with a square matrix in a way that proves to be useful in this context.

Let us start with the linear equations $\boldsymbol{A}\boldsymbol{x} = \boldsymbol{h}$ and write them out in full for the case of two equations in two unknowns:

$$\begin{aligned} a_{11}x_1 + a_{12}x_2 &= h_1, \\ a_{21}x_1 + a_{22}x_2 &= h_2. \end{aligned}$$

Multiplying the first of these by a_{21} and the second by a_{11} and subtracting, we find

$$(a_{11}a_{22} - a_{12}a_{21})x_2 = -a_{21}h_1 + a_{11}h_2$$

and similarly

$$(a_{11}a_{22} - a_{12}a_{21})x_1 = a_{22}h_1 - a_{12}h_2.$$

Both solutions involve the quantity $a_{11}a_{22} - a_{12}a_{21}$ which depends only upon the elements of the coefficient matrix and is obtained from them as a sort of cross-difference. If this quantity is nonzero, we can immediately get explicit solutions for x_1 and x_2. The expression has (rather obviously) two terms, one positive and one negative, and each term is the product of two elements, one from each row and one from each column of the coefficient matrix.

Now repeat all this for the case of three equations in three unknowns. The reader should set out the algebra; it will be found again that the expression for each of the unknowns involves a single quantity given by

$$\begin{aligned} a_{11}a_{22}a_{33} &- a_{11}a_{23}a_{32} \\ &+ a_{12}a_{23}a_{31} - a_{12}a_{21}a_{33} \\ &+ a_{13}a_{21}a_{32} - a_{13}a_{22}a_{31} \end{aligned}$$

which must be nonzero for the solution to be completed. It has six terms, three positive and three negative, and each term is the product of three elements, one from each row and one from each column of the matrix A.

2.2 Definition of a determinant

The direction of generalization is clear. We write the *determinant* of the square $p \times p$ matrix A as $|A|$ and define it as follows:

1. $|A|$, a scalar quantity, is the sum of $p!$ terms, where $p!$ means $p(p-1)(p-2) \cdot \cdots \cdot 2 \cdot 1$. Each term is a product of p elements, one from each row and one from each column.

2. Half the terms in $|A|$ are positive and half negative, according to a rule to be developed below.

2.3 Permutations

If A is $p \times p$, we can write each of the terms in the determinant in the form $a_{1i}a_{2j} \ldots a_{pk}$ where the row suffixes are in their natural order and the column suffixes $(ij \ldots k)$ are a rearrangement or *permutation* of the numbers $1, 2, \ldots, p$. There are just $p!$ such permutations (for there are p possible choices for the first column suffix, then $p-1$ for the second, then $p-2$ for the third, and so on) and these between them account for all the $p!$ terms in the determinant.

A particularly simple kind of permutation is a *transposition* (the reader must suppress the matrix meaning of this term for the time being) which consists in the interchange of two items. One permutation can be obtained from another by imposing successively one or more transpositions; for example, to get from 1234 to 2314, we could interchange first the second and third items (getting 1324) and then the first and third (getting 2314 as required). The number of transpositions needed to get to a particular permutation $ij \ldots k$ starting from the natural order $12 \ldots p$ is not fixed; the change can be made in many different ways, but the number of transpositions involved turns out to be always even or always odd, depending on the target permutation concerned. This can most easily be seen by a rather neat proof which takes $x_1, x_2, \ldots, x_p$ to be p different numbers and considers the product $\Pi_{i \neq j}(x_j - x_i)$. Suppose we apply a permutation to the suffixes in this product. Then at most we shall change the signs of some of the terms in the product and so multiply its value by either $+1$ or -1. Now consider a single transposition, which interchanges l and m, say. Factors that do not involve x_l or x_m are unchanged; so are pairs of factors like $(x_n - x_l)(x_n - x_m)$; the single factor $x_l - x_m$ changes its sign; consequently the whole product is multiplied by -1. But if the whole permutation can be obtained by making successively each of a number of transpositions, the effect on the product as a whole must be the same whatever the sequence, and so the number of multiplications by -1 must be always even or always odd.

2.4 Definition of a determinant—continued

By examining their makeup as sequences of transpositions, we can divide all permutations into two classes, even and odd. This enables us to complete our definition of a determinant, since each term in the expansion corresponds to a permutation as pointed out above; we take the terms corresponding to even permutations with positive signs and those corresponding to odd permutations with negative signs. Note that this definition is actually symmetric as between rows and columns: whenever we have a result which is true for the rows of a determinant, there will be a similar result for the columns.

The reader should check that the definition works for the 2×2 and 3×3 matrices in Section 2.1. Fairly obviously, the permutation 12 is even (with 0 transpositions) and 21 is odd (with 1 transposition). 123 and its cyclic variants 231 and 312 are even; 132, 213, and 321 are odd.

One or two important findings follow immediately from this definition. For a start, $|A| = |A^{\mathrm{T}}|$, from the symmetry of the definition as between rows and columns. Secondly, consider the effect on $|A|$ of interchanging two columns. Every term in the expansion of $|A|$ corresponds to a permutation and the interchange applies a single transposition to each of these permutations. It thus changes all even permutations to odd and all odd permutations to even, and so simply changes the sign of the determinant. Now, suppose A has two identical columns; then interchanging them has no effect. But the only number which is not altered when its sign is changed is zero, and so we have proved that the determinant of a matrix with two identical columns (or two identical rows) vanishes.

2.5 Minors and cofactors

If A is $p \times p$ and we focus on a single row of A, say the ith, then $|A|$ is a linear combination of the elements of this row. Separating out all the terms involving a_{i1}, all those involving a_{i2}, . . . and so on, we can write

$$|A| = \alpha_{i1}a_{i1} + \alpha_{i2}a_{i2} + \cdots + \alpha_{ip}a_{ip}$$

(remember that each term in the expansion of the determinant contains just one element from this row). If we enumerate the individual terms in the coefficient α_{ij} we find that there are $(p - 1)!$ of them and that each one is the product of $p - 1$ elements from A, one from each row except the ith and one from each column except the jth. Half of these terms are positive and half are negative, and these signs are such that α_{ij} is just plus or minus the determinant of a matrix which is obtained from A by deleting the ith row and the jth column (the sign is actually $(-1)^{i+j}$). Such a determinant is called a *minor* and when the appropriate sign is attached it is called a *cofactor*: in particular, the minor or cofactor of the element which it multiplies in the expression above. Both definitions can be extended to apply to the determinants of square submatrices obtained from A by deleting any selection of rows and columns. If the deleted rows and columns are in the same positions, so that the resulting submatrix's main diagonal is part of that of A itself, the minor is called a *principal minor*. Note that A_{ij} is the more usual notation for α_{ij}, but in this book we reserve capital letters for matrices.

2.6 Further properties of determinants

Several results follow from the expansion of a determinant in terms of the elements of a single row. First, it is clear that multiplying each element of one row by a constant multiplies the determinant by the same constant. This means that $|kA| = k^p|A|$, and more generally that, if $R = r^{\mathrm{d}}$ is a diagonal matrix, then $|AR| = \Pi r_i \cdot |A| = |A| \cdot |R|$. This is a special case of a more general result. Next, it follows that the determinant of a matrix vanishes, not merely when two rows are equal, but when two rows have elements which are in constant proportion.

Consider the expression

$$\alpha_{i1}a_{j1} + \alpha_{i2}a_{j2} + \cdots + \alpha_{ip}a_{jp}$$

with $j \neq i$. This is the determinant of a matrix which is obtained by replacing the ith row of A by its jth row. But this matrix has two identical rows, and consequently this *expansion by alien cofactors* vanishes. From this we can see that

$$\alpha_{i1}(a_{i1} + ka_{j1}) + \alpha_{i2}(a_{i2} + ka_{j2}) + \cdots + \alpha_{ip}(a_{ip} + ka_{jp}) = |A|$$

if $j \neq i$, so that adding any multiple of one row to another does not affect the determinant.

2.7 The determinant of a product

A most important result, of which we have already had a special case, states that, if A and B are both $p \times p$, then $|AB| = |BA| = |A||B|$. There are various 'clever' proofs of this, which require rather more preliminary apparatus than is worth while here; it can also be proved quite easily by brute force, by enumerating the various terms involved.

2.8 Determinants and simultaneous equations

Consider the pair of equations, written as a vector equation,

$$\begin{bmatrix} 2 & 1 \\ 4 & 3 \end{bmatrix} \begin{bmatrix} x_1 \\ x_2 \end{bmatrix} = \begin{bmatrix} 8 \\ 12 \end{bmatrix}.$$

Subtracting twice the first from the second, we find that

$$0x_1 + 1x_2 = -4$$

so that $x_2 = -4$. The first equation now gives

$$2x_1 = 8 - x_2 = 12$$

so that $x_1 = 6$, and this is checked by the second equation

$$4 \cdot 6 + 3 \cdot (-4) = 12.$$

It may be clear (we shall prove it later) that this solution to the equations is *unique*; no other pair of numbers will satisfy them. The determinant of the coefficient matrix is $2 \cdot 3 - 1 \cdot 4 = 2$.

Now consider the equation pair

$$\begin{bmatrix} 2 & 1 \\ 4 & 2 \end{bmatrix} \begin{bmatrix} x_1 \\ x_2 \end{bmatrix} = \begin{bmatrix} 8 \\ 12 \end{bmatrix}.$$

If we try the same trick as before, subtracting twice the first equation from the second so as to get rid of x_1, we get

$$0x_1 + 0x_2 = -4,$$

a contradiction; there are no values of x_1 and x_2 which satisfy these equations. On the other hand, if we try

$$\begin{bmatrix} 2 & 1 \\ 4 & 2 \end{bmatrix} \begin{bmatrix} x_1 \\ x_2 \end{bmatrix} = \begin{bmatrix} 6 \\ 12 \end{bmatrix},$$

this leads us to the relation

$$0x_1 + 0x_2 = 0,$$

which is true for any values of x_1 and x_2. If we, quite arbitrarily, take $x_2 = 4$, then the first equation gives

$$2x_1 + 1 \cdot 4 = 6,$$

so that $x_1 = 1$ and the second equation is satisfied; if we take $x_2 = 2$, then $x_1 = 2$ again satisfies both the equations. There are in fact an infinite number of possible solutions to the equations.

The key to the situations in the last two examples turns out to be the determinant of the coefficient matrix, which is $2 \cdot 2 - 4 \cdot 1 = 0$. As we shall show in more detail later, when the determinant of the coefficient matrix is nonzero, the equations always have a unique solution. When the determinant vanishes, there may be no solution to the equations at all; when there is one solution, there will also be an infinite number of other ones.

Examples **2**

2.1 The expansion of a determinant in terms of a single row or column is a very practical way of evaluating small determinants, up to perhaps 4×4 or so. We have

$$\begin{bmatrix} a_{11} & a_{12} \\ a_{21} & a_{22} \end{bmatrix} = a_{11}a_{22} - a_{12}a_{21}$$

and

$$\begin{bmatrix} a_{11} & a_{12} & a_{14} \\ a_{21} & a_{22} & a_{23} \\ a_{31} & a_{32} & a_{33} \end{bmatrix}$$

$$= a_{11}\begin{bmatrix} a_{22} & a_{23} \\ a_{32} & a_{33} \end{bmatrix} - a_{12}\begin{bmatrix} a_{21} & a_{23} \\ a_{31} & a_{33} \end{bmatrix} + a_{13}\begin{bmatrix} a_{21} & a_{22} \\ a_{31} & a_{32} \end{bmatrix},$$

the 2×2 determinants being the minors of the elements in the first row. The reader should evaluate the determinants of the following matrices to see how the procedure works:

$$\begin{bmatrix} 4 & 2 \\ 1 & 7 \end{bmatrix}, \quad \begin{bmatrix} 3 & 1 & 5 \\ 4 & 8 & 6 \\ 2 & 4 & 2 \end{bmatrix}, \quad \begin{bmatrix} 1 & 3 & 4 & 2 \\ 2 & -1 & 1 & 3 \\ 2 & 5 & -2 & 4 \\ -1 & 4 & 3 & -1 \end{bmatrix},$$

2.2 If most of the elements of a row (or column) of a matrix are zero, this is an obvious row (or column) to choose for expanding the determinant.

Show in this way that the determinant of a diagonal matrix is equal to the product of the elements on the main diagonal (a result we used in Section 2.6), and that the same is true of a triangular matrix.

2.3 We can *create* zeros in a chosen row of a matrix by adding or subtracting suitable multiples of other rows—we know that this does not alter the value of the determinant (see Section 2.6). The reader should try subtracting twice the third row from the second row in the 3×3 matrix of example 1; and adding the second row to the fourth row and then subtracting the first row from the result in the 4×4 matrix in the same example. Practical methods for evaluating larger determinants are based on systematic use of this procedure

2.4 If

$$A = \begin{bmatrix} 1 & 4 & 3 \\ 2 & 1 & 5 \\ 3 & 2 & 4 \end{bmatrix} \quad \text{and} \quad B = \begin{bmatrix} 2 & 4 & 5 \\ 3 & 1 & 4 \\ 2 & 2 & 3 \end{bmatrix},$$

work out AB and BA and verify that $|AB| = |BA| = |A||B|$

2.5 Let A be the partitioned matrix $\begin{bmatrix} I & 0 \\ B & C \end{bmatrix}$.

Show that $|A| = |C|$.

3
Inverse matrices

3.1 Introduction

We can now return to the subject of matrix division, and we again use the solution of simultaneous equations to motivate our definition. The single equation

$$ax = h$$

with $a \neq 0$ has the solution $x = h/a$, or as we can write it, $x = a^{-1}h$. Consider now p equations in p unknowns and write them in the form

$$A\boldsymbol{x} = \boldsymbol{h}$$

with the coefficients in a square $p \times p$ matrix A. Assuming for the moment that some vector $\boldsymbol{x}$ exists which satisfies the equations, we would like to find that $\boldsymbol{x}$ could be written as $\boldsymbol{x} = B\boldsymbol{h}$, where B will be a matrix which we can call the *inverse* of A (the reciprocal of A would be a possible alternative name) and can write as A^{-1}. This will work if we can find a $p \times p$ matrix B which is such that $BA = I$, a unit matrix; for then we would simply have to multiply both sides of the equations by B to obtain the solutions. Note that we only deal with square matrices at this stage, so that we have as many equations as there are unknowns.

3.2 Definition of the inverse matrix

Consider first the $p \times p$ matrix C which is such that

$$c_{ij} = \alpha_{ji}$$

where α_{ji} denotes the cofactor in $|A|$ of a_{ji} (note the implied transposition). C is called the *adjugate* or *adjoint* matrix of A and we write $C = \text{adj}\, A$. Now the (i, j)th element of the product CA is

$$\Sigma_k c_{ik} a_{kj} = \Sigma_k \alpha_{ki} a_{kj}.$$

If $i = j$, this sum of products is equal to $|A|$, being the expansion of the determinant by its ith column; if $i \neq j$, it is an expansion by alien cofactors and is equal to zero. Thus C is a scalar matrix equal to $|A|I$. Provided that $|A| \neq 0$ we can define A^{-1} to be $(1/|A|)\text{adj}\, A$ and this matrix will have just the property $A^{-1}A = I$ that we require.

This of course fails if $|A| = 0$. In this case A is called *singular*. Since $|A^{-1}||A| = |A^{-1}A| = |I| = 1$, there can be no inverse of a singular matrix (any more than there can be a reciprocal of zero), though we shall devote Chapter 5 to something very much like one. A matrix that is not singular is sometimes called *regular*, but *nonsingular* is more common. We have just shown that every nonsingular matrix possesses an inverse.

3.3 Properties of the inverse

If, as above, $C = \text{adj}\, A$ then the (i,j)th element of the product AC is $\sum_k a_{ik}\alpha_{jk}$, and it appears at once that $AC = |A|I$. With our definition of the inverse it follows that when A^{-1} exists then $AA^{-1} = I$; in other words, that A and its inverse commute. We can also show that A has only a single inverse; for if $AB = I = AC$ then $A^{-1}AB = A^{-1}AC$, and so $B = C$. It follows from this that when A is nonsingular, the equations $A\boldsymbol{x} = \boldsymbol{h}$ always have a single, unique solution.

We leave the reader to prove some other simple properties of the inverse. Assuming that the appropriate inverses exist:

1. $(A^{-1})^{-1} = A$.
2. $(kA)^{-1} = (1/k)A^{-1}$ (provided $k \neq 0$).
3. $(A^{\mathsf{T}})^{-1} = (A^{-1})^{\mathsf{T}}$.
4. $(AB)^{-1} = B^{-1}A^{-1}$
   (compare the similar 'reversal' rule for the transpose of a product of two matrices).
5. $|A^{-1}| = 1/|A|$.
6. The inverse of a symmetric matrix is itself symmetric.
7. The inverse of the diagonal matrix $\boldsymbol{x}^{\mathrm{d}}$ with

$$\boldsymbol{x}^{\mathsf{T}} = [x_1 \quad x_2 \quad \ldots \quad x_p]$$

is $\boldsymbol{y}^{\mathrm{d}}$ with

$$\boldsymbol{y}^{\mathsf{T}} = [1/x_1 \quad 1/x_2 \quad \ldots \quad 1/x_p].$$

This may be written $\boldsymbol{x}^{-\mathrm{d}}$.

Inverting (forming the inverse of) a matrix of any size involves a fair amount of arithmetic, which we shall discuss briefly in Chapter 8. Even with efficient numerical methods, the amount of work goes up as the cube of the size of the matrices concerned. Working straight from the definition by calculating the adjugate and the determinant is really only worth while for $p = 2$: it is worth remembering that

$$\begin{bmatrix} a_{11} & a_{12} \\ a_{21} & a_{22} \end{bmatrix}^{-1} = \begin{bmatrix} a_{22} & -a_{12} \\ -a_{21} & a_{11} \end{bmatrix} / (a_{11}a_{22} - a_{12}a_{21}).$$

3.4 The inverse of a partitioned matrix

Suppose we wish to invert the $p \times p$ matrix A and that we partition the rows and columns in the same way so that

$$A = \begin{bmatrix} A_{11} & A_{12} \\ A_{21} & A_{22} \end{bmatrix}$$

with A_{11} and A_{22} both square, of sizes $r \times r$ and $(p-r) \times (p-r)$ say. Suppose we partition the inverse in the same way and write

$$A^{-1} = \begin{bmatrix} A^{11} & A^{12} \\ A^{21} & A^{22} \end{bmatrix}.$$

Then we can treat the submatrices as if they were elements, and because $AA^{-1} = I$ we have

$$\begin{aligned} A_{11}A^{11} + A_{12}A^{21} &= I_r, \\ A_{11}A^{12} + A_{12}A^{22} &= 0, \\ A_{21}A^{11} + A_{22}A^{21} &= 0, \\ A_{21}A^{12} + A_{22}A^{22} &= I_{p-r}. \end{aligned}$$

From the third of these

$$A^{21} = -A_{22}{}^{-1}A_{21}A^{11}$$

assuming that A_{22} is nonsingular and has an inverse; and then substituting in the first of the four equations gives

$$(A_{11} - A_{12}A_{22}{}^{-1}A_{21})A^{11} = I,$$

and hence

$$A^{11} = (A_{11} - A_{12}A_{22}{}^{-1}A_{21})^{-1},$$

again assuming that the inverse exists. In the same way, from $A^{-1}A = I$ we get the four relations

$$\begin{aligned} A^{11}A_{11} + A^{12}A_{21} &= I, \\ A^{11}A_{12} + A^{12}A_{22} &= 0, \\ A^{21}A_{11} + A^{22}A_{21} &= 0, \\ A^{21}A_{12} + A^{22}A_{22} &= I. \end{aligned}$$

The second of these gives

$$A^{12} = -A^{11}A_{12}A_{22}{}^{-1}$$

and from the fourth we have

$$A^{22} = A_{22}{}^{-1} - A^{21}A_{12}A_{22}{}^{-1}.$$

We now see that we can invert the whole matrix if we can find the inverses of the $(p-r)\times(p-r)$ matrix A_{22} and the $r\times r$ matrix $A_{11}-A_{12}A_{22}^{-1}A_{21}$. This may be very practical if A_{22} is easy to invert—diagonal for instance—and if the other matrix is small, perhaps 1×1 or 2×2.

Suppose for example that we know the inverse of a matrix (which we write as A_{22}) and that we want to know that of a larger matrix formed by adding a row and column to it. The new matrix to be inverted is merely 1×1, and all that we need is the reciprocal of an ordinary number. This provides a very practical way of inverting a general square matrix by starting in one corner and building up the whole matrix by adding one row and column at a time. The process is slightly simpler when A is symmetric (as is the case in statistical applications) since then $A^{21}=(A^{12})^{\mathrm{T}}$ and some of the work can be avoided.

In much the same way, we can find the inverse of a submatrix of a matrix we have already inverted, in particular one formed by removing a single row and column. In the notation we have been using, if we know the inverse of A and can readily reinvert A^{11}, we can get the inverse of A_{22} (which will in general be different from A^{22}). We have in fact

$$A_{22}^{-1}=A^{22}+A^{21}A_{12}A_{22}^{-1}.$$

But

$$\begin{aligned}A^{21}A_{12}A_{22}^{-1}&=A^{21}(A^{11})^{-1}A^{11}A_{12}A_{22}^{-1}\\&=-A^{21}(A^{11})^{-1}A^{12}\end{aligned}$$

so that A_{22}^{-1} can be calculated from $(A^{11})^{-1}$ and the submatrices of the original inverse.

These methods have direct statistical applications in multiple regression calculations. Suppose that S is the symmetric matrix of sums of squares and products of a set of x-variables and invert it to form S^{-1}. Now add to S a further row and column relating to a y-variate and find the inverse of the enlarged matrix by the method we have described. If we write the enlarged inverse as

$$\begin{bmatrix}P & \boldsymbol{q}\\ \boldsymbol{q}^{\mathrm{T}} & r\end{bmatrix}$$

we find that the vector $\boldsymbol{q}$ contains the regression coefficients of y on the x's, all times $-1/r$, while $1/r$ itself is equal to the residual sum of squares of y after allowing for regression on the x's.

Once we have identified these elements of the enlarged inverse, we can arrange the calculations in a different way. Form the enlarged matrix of sums of squares and products, now with the sums of squares and products of the y-variate forming the first row and column, and invert this. The regression coefficients and the residual sum of squares can now be obtained from the elements of the first row or column of the inverse. However, it is now straightforward to add new rows and columns to the inverse matrix which correspond to new x-variables, or to remove rows and columns which are already there, estimating in this way the regressions of y on different selections of the x-variables. This method is used by many of the standard computer programs for doing multiple regression.

3.5 Orthogonal matrices

In Section 1.7 we defined a square matrix X to be orthogonal if every pair of columns, $\boldsymbol{x}_i$ and $\boldsymbol{x}_j$ say, were orthonormal, i.e. $\boldsymbol{x}_i^T\boldsymbol{x}_i = 1$, $\boldsymbol{x}_j^T\boldsymbol{x}_j = 1$, $\boldsymbol{x}_i^T\boldsymbol{x}_j = 0$, so that $X^TX = I$. It is immediate that $X^T = X^{-1}$, so that the inverse of an orthogonal matrix is just its transpose. Moreover it follows that $XX^T = I$, showing that the rows of an orthogonal matrix, as well as the columns, consist of a set of orthonormal vectors. Note that $|X|^2 = 1$, so that the determinant of an orthogonal matrix is equal to ± 1.

Examples 3

3.1 From example 1.2 we know that if

$$A = \begin{bmatrix} 4 & 8 & 4 \\ 8 & 25 & 11 \\ 4 & 11 & 30 \end{bmatrix} \quad \text{and} \quad U = \begin{bmatrix} 2 & 4 & 2 \\ 0 & 3 & 1 \\ 0 & 0 & 5 \end{bmatrix}$$

then $A = U^TU$. By setting up and solving the equations $U\boldsymbol{x} = \boldsymbol{h}$ with $\boldsymbol{h}$ successively equal to $[1, 0, 0]^T$, $[0, 1, 0]^T$, and $[0, 0, 1]^T$, find the inverse of U and from this derive the inverse of A.

3.2 If the numbers of cases in the cells of a two-way table are

4	7	5
3	5	2
2	2	6
4	8	7,

the least-squares equations for fitting a main-effects model have a coefficient matrix given by

$$\begin{bmatrix} 55 & 13 & 22 & 20 & 16 & 10 & 10 & 19 \\ 13 & 13 & 0 & 0 & 4 & 3 & 2 & 4 \\ 22 & 0 & 22 & 0 & 7 & 5 & 2 & 8 \\ 20 & 0 & 0 & 20 & 5 & 2 & 6 & 7 \\ 16 & 4 & 7 & 5 & 16 & 0 & 0 & 0 \\ 10 & 3 & 5 & 2 & 0 & 10 & 0 & 0 \\ 10 & 2 & 2 & 6 & 0 & 0 & 10 & 0 \\ 19 & 4 & 8 & 7 & 0 & 0 & 0 & 19 \end{bmatrix}.$$

This matrix is singular and has no inverse, but a nonsingular matrix can be produced by crossing out the first and second rows and columns. Invert this matrix by exploiting the simple submatrices on the main diagonal.

[The matrix to be inverted is

$$X = \begin{bmatrix} 22 & 0 & 7 & 5 & 2 & 8 \\ 0 & 20 & 5 & 2 & 6 & 7 \\ 7 & 5 & 16 & 0 & 0 & 0 \\ 5 & 2 & 0 & 10 & 0 & 0 \\ 2 & 6 & 0 & 0 & 10 & 0 \\ 8 & 7 & 0 & 0 & 0 & 19 \end{bmatrix}.$$

Writing $X = \begin{bmatrix} A & B \\ B & C \end{bmatrix}$ and $X^{-1} = \begin{bmatrix} P & Q \\ Q & R \end{bmatrix}$,

we have

$$\begin{aligned} B^{\mathrm{T}}P + CQ^{\mathrm{T}} &= 0, & Q^{\mathrm{T}} &= -C^{-1}B^{\mathrm{T}}P, \\ AP + BQ^{\mathrm{T}} &= I, & P &= (A - BC^{-1}B^{\mathrm{T}})^{-1}, \\ C^{-1} &= [\tfrac{1}{16} \ \ \tfrac{1}{10} \ \ \tfrac{1}{10} \ \ \tfrac{1}{19}]^{\mathrm{d}}, \end{aligned}$$

so $BC^{-1}B^{\mathrm{T}}$ is the weighted SSP matrix of the rows of B:

$$\begin{bmatrix} 9.3309 & 7.3349 \\ & 8.1414 \end{bmatrix}.$$

Thus $P = \begin{bmatrix} 12.6691 & -7.3349 \\ & 11.8586 \end{bmatrix}^{-1} = \begin{bmatrix} 11.8586 & 7.3349 \\ & 12.6691 \end{bmatrix} / 96.4367$

$$= \begin{bmatrix} 0.122967 & 0.076059 \\ & 0.131372 \end{bmatrix}.$$

Also $CB^{\mathrm{T}} = \begin{bmatrix} 0.437500 & 0.312500 \\ 0.500000 & 0.200000 \\ 0.200000 & 0.600000 \\ 0.421053 & 0.368421 \end{bmatrix}$ so $Q^{\mathrm{T}} = \begin{bmatrix} -0.077567 & -0.074330 \\ -0.076695 & -0.064304 \\ -0.070229 & -0.094035 \\ -0.079797 & -0.080425 \end{bmatrix}$.

Finally $B^{\mathrm{T}}Q + CR = I$, $R = C^{-1} - C^{-1}B^{\mathrm{T}}Q$, and so

$$R = \begin{bmatrix} 0.119664 & 0.053650 & 0.060111 & 0.060044 \\ & 0.151208 & 0.053922 & 0.055984 \\ & & 0.170467 & 0.064214 \\ & & & 0.115861 \end{bmatrix}.$$

Note that the lower halves of the symmetric matrices have been omitted.]

3.3 Verify that the following rows make up sets of orthogonal contrasts:

(a)
$$\begin{bmatrix} 1 & 1 & 1 & 1 & 1 & 1 & 1 & 1 \\ 1 & 1 & 1 & 1 & -1 & -1 & -1 & -1 \\ 1 & 1 & -1 & -1 & 1 & 1 & -1 & -1 \\ 1 & 1 & -1 & -1 & -1 & -1 & 1 & 1 \\ 1 & -1 & 1 & -1 & 1 & -1 & 1 & -1 \\ 1 & -1 & 1 & -1 & -1 & 1 & -1 & 1 \\ 1 & -1 & -1 & 1 & 1 & -1 & -1 & 1 \\ 1 & -1 & -1 & 1 & -1 & 1 & 1 & -1 \end{bmatrix},$$

(b)
$$\begin{bmatrix} 1 & 1 & 1 & 1 & 1 \\ 4 & -1 & -1 & -1 & -1 \\ 0 & 3 & -1 & -1 & -1 \\ 0 & 0 & 2 & -1 & -1 \\ 0 & 0 & 0 & 1 & -1 \end{bmatrix}.$$

(c) the values of the following polynomials for $t = 1, 2, 3, 4, 5$:

$$t^0, \quad t^1 - 3, \quad (t-3)^2 - 2,$$
$$5(t-3)^3 - 17(t-3),$$
$$35(t-3)^4 - 155(t-3)^2 + 72.$$

3.4 The SSP matrix of a y-variate and three x's is

$$\begin{array}{c} \\ y \\ x_1 \\ x_2 \\ x_3 \end{array} \begin{array}{c} \begin{array}{cccc} y & x_1 & x_2 & x_3 \end{array} \\ \begin{bmatrix} 9.9868 & -0.2646 & 4.4459 & 6.8938 \\ & 24.1277 & 1.1228 & 2.4337 \\ & & 2.7411 & 3.6366 \\ & & & 10.0771 \end{bmatrix} \end{array}.$$

The matrix consisting of the first three rows and columns has been inverted, giving

$$\begin{array}{c} \\ y \\ x_1 \\ x_2 \end{array} \begin{array}{c} \begin{array}{ccc} \quad y \quad & \quad x_1 \quad & \quad x_2 \quad \end{array} \\ \begin{bmatrix} 0.38580 & 0.03400 & -0.63967 \\ & 0.04525 & -0.07368 \\ & & 1.43251 \end{bmatrix} \end{array}.$$

(a) Interpret this inverse in terms of the multiple regression of y on x_1 and x_2.

(b) What is the effect of

1. Adding x_3 to the regression?
2. Then removing x_1 from the regression?

[(a) The regression coefficients of y on x_1 and x_2 are

$$b_1 = -0.0881, \qquad b_2 = 1.6580,$$

and the residual sum of squares is $1/0.38580 = 2.5920$.

(b) (1) As in example 2 above, let $\begin{bmatrix} A & B \\ B^{\mathsf{T}} & C \end{bmatrix}^{-1} = \begin{bmatrix} P & Q \\ Q^{\mathsf{T}} & R \end{bmatrix}$.

We know A^{-1} and we have

$$B^{\mathsf{T}} = [6.8938\ 2.4337\ 3.6366], \qquad C = [10.0771].$$

Then $(A^{-1}B)^{\mathsf{T}} = [0.41614, 0.07656, 0.62040]$
and $B^{\mathsf{T}}A^{-1} = 5.31125$
so $R = 1/4.76585 = 0.20983$.
Then $Q^{\mathsf{T}} = -(A^{-1}BR)^{\mathsf{T}} = [-0.08732, -0.01606, -0.13017]$
and $P = A^{-1} + A^{-1}BQ$ so that the 4×4 inverse matrix is

$$\begin{bmatrix} 0.42214 & 0.04068 & -0.58550 & -0.08732 \\ & 0.04648 & -0.06371 & -0.01606 \\ & & 1.51326 & -0.13017 \\ & & & 0.20982 \end{bmatrix}$$

Thus the new regression coefficients are

$$b_1 = -0.0964, \qquad b_2 = 1.3870, \qquad b_3 = 0.2068$$

and the residual sum of squares is now $1/0.42214 = 2.3689$.

(2) This calculation needs a little care in selecting the required elements of the matrices. We find that the 3 × 3 inverse is given by

$$\begin{bmatrix} 0.42214 & -0.58550 & -0.08732 \\ & 1.51326 & -0.13017 \\ & & 0.20982 \end{bmatrix}$$

$$- \begin{bmatrix} 0.04068 \\ -0.06371 \\ -0.01606 \end{bmatrix} (1/0.04648)[0.04068 \quad -0.06371 \quad -0.01606]$$

$$= \begin{bmatrix} 0.38654 & -0.52974 & -0.07326 \\ & 1.42593 & -0.15218 \\ & & 0.20427 \end{bmatrix}.$$

The regression coefficients are now

$$b_2 = 1.3705, \qquad b_3 = 0.1895$$

and the regression sum of squares is 1/0.38654 = 2.5871].

3.5 Let A be nonsingular and partitioned as $\begin{bmatrix} A_{11} & A_{12} \\ A_{21} & A_{22} \end{bmatrix}$;

then $|A| = |A_{22}|\,|A_{11} - A_{12}A_{22}^{-1}A_{21}|$.

[Consider the matrix $B = \begin{bmatrix} I & 0 \\ -A_{22}^{-1}A_{21} & A_{22}^{-1} \end{bmatrix}$;

then $AB = \begin{bmatrix} A_{11} - A_{12}A_{22}^{-1}A_{21} & A_{12}A_{22}^{-1} \\ 0 & I \end{bmatrix}$.

It follows from example 2.5 that $|AB| = |A_{11} - A_{12}A_{22}^{-1}A_{21}|$, but $|B| = |A_{22}^{-1}|$ and the required result follows.]

4
Linear dependence and rank

4.1 Definitions

Suppose we have a set of vectors $\boldsymbol{x}_1, \boldsymbol{x}_2, \ldots, \boldsymbol{x}_p$, all with the same number of elements, and a set of scalar coeficients $c_1, c_2, \ldots, c_p$, not all of them zero, which are such that

$$c_1\boldsymbol{x}_1 + c_2\boldsymbol{x}_2 + \cdots + c_p\boldsymbol{x}_p = \mathbf{0},$$

where the right-hand side is a vector of zeros. We call this a *linear relationship* between the x-vectors and say that the vectors are *linearly dependent.* Notice that if we choose any one of the c_i which is not zero, the corresponding $\boldsymbol{x}_i$ can be written as a linear combination of the other $\boldsymbol{x}$'s; and conversely, that if one of the $\boldsymbol{x}$-vectors can be written as a linear combination of the others, then the whole lot taken together are linearly dependent. A set of vectors $\boldsymbol{x}_i$ for which no zero-valued linear combination can be formed (apart from the trivial one with all the coefficients zero) is called linearly independent.

The *rank* of a matrix A, which we write as $\mathrm{r}(A)$, is simply the size of the largest square submatrix of A which is not singular. If A is $p \times p$ and nonsingular, then $\mathrm{r}(A) = p$. If A is $m \times n$, $\mathrm{r}(A)$ cannot exceed the smaller of m and n. For a square matrix, the size minus the rank is called the *nullity* (so that the nullity of a nonsingular square matrix is 0). The apparently quite distinct notions of linear dependence and rank are actually closely connected, as we proceed to show.

4.2 Matrices with linearly dependent rows or columns

We can easily show that a square matrix with linearly dependent rows (or columns) is singular. Suppose that A is $p \times p$ with the vectors $\boldsymbol{a}_1, \boldsymbol{a}_2, \ldots, \boldsymbol{a}_p$ as rows and that

$$c_1\boldsymbol{a}_1 + c_2\boldsymbol{a}_2 + \cdots + c_p\boldsymbol{a}_p = \mathbf{0},$$

with not all the c's zero. Suppose in fact that $c_i \neq 0$; then we can multiply the ith row of A by c_i (thereby multiplying $|A|$ by the same factor), and then we can add to this row c_j times the jth row, where j refers successively to all the other rows of the matrix (leaving the determinant unaltered throughout). The result of all this is a matrix with a whole row consisting of zeros. Expanding by this row shows that this matrix has a determinant of zero. Accordingly $c_i\,|A| = 0$, and consequently, since $c_i \neq 0$, we have $|A| = 0$, $\boldsymbol{A}$ is singular, and $\mathrm{r}(\boldsymbol{A}) < p$.

Now suppose that A is rectangular, $m \times n$ say, with $n > m$, so that in any case $\mathrm{r}(A)$ cannot exceed m. If the m rows of A are linearly dependent, this means that there is a linear relationship between the rows with coefficients $c_1, c_2 \ldots, c_m$ not all zero; but this equality between vectors is just shorthand for n equalities, all with the same coefficients, relating to the elements in a single column. The same equality must apply also to any set of subvectors obtained by deleting all but m of the columns of A. As a result, any $m \times m$ submatrix of A must have linearly dependent rows and so must be singular. It follows that the rank of A must be less than m. As a sort of converse of this, note that the columns of A are always linearly dependent if $n > m$. If $\mathrm{r}(A) = m$, we can find a set of m columns making up a nonsingular submatrix, and we can use this as the coefficient matrix of a set of linear equations. By putting each of the other columns in turn on the right-hand side and solving the equations (which we can always do), we can express each of these other columns in terms of the original set of m. If $\mathrm{r}(A) < m$, there will already be a linear relationship between m of the columns, and we can include the other columns in this by giving them zero coefficients.

4.3 Reduction to canonical form

We know that we can permute the rows (or columns) of a square matrix without altering the absolute value of the determinant. We can also add any multiple of one row (or column) to another without the determinant changing. If we multiply all the elements of a single row (or column) by a nonzero constant, we multiply the value of the determinant by the same constant. These three operations on a square matrix are called *elementary operations*. We now see that if we perform any elementary operation, or sequence of elementary operations, on a general square matrix, we do not cause the determinant of any submatrix to change from nonzero to zero, and consequently we do not change the rank.

It should be noted that the elementary operations can all be effected by multiplication by a suitable matrix, on the left to affect rows, on the right to affect columns. The required matrix is easily formed by performing the required elementary operation on the rows or columns of a unit matrix. Thus to make a given interchange of rows or columns, the required matrix is a unit matrix with the same interchange done on it. The matrix which adds k times the ith column to the jth column is a unit matrix with k inserted as the (i, j)th element. The matrix which multiplies the ith column by k is a unit matrix with the ith diagonal element replaced by k. It is important to note that all these elementary operation matrices are nonsingular and so possess inverses, and that these inverse matrices correspond to other elementary operations which merely undo the original ones.

Now consider a general matrix $A \neq 0$. Let A be $m \times n$ and arrange by permuting the rows and columns as necessary that the top left-hand element $a_{11} \neq 0$. If we now subtract a_{12}/a_{11} times the first row from the second, we produce a zero in the (1, 2) element. In the same way we can produce zeros all the way down the first column; and we can do the same thing along the first row. Finally, multiply the first row by $1/a_{11}$. All the time we have been performing elementary operations, and the result is that we end up with a matrix which we can write in partitioned form as

$$\begin{bmatrix} 1 & \mathbf{0}^{\mathrm{T}} \\ \mathbf{0} & A_1 \end{bmatrix},$$

which has the same rank as A.

Now the matrix A_1 in the bottom right-hand corner may be a zero matrix. If it is not, we can repeat the process, getting a nonzero element into the (2, 2) position and using it to produce zeros along the second row and column—all this without changing the elements in the first row and column. We reach a matrix, still with the same rank as A, which can be written as

$$\begin{bmatrix} I_2 & 0 \\ 0 & A_2 \end{bmatrix}.$$

If A_2 is not a zero matrix, we can repeat the process again. The whole thing comes to a halt either when we run out of rows or columns, or else when one of the A_i matrices turns out to have no nonzero elements. The final outcome is a matrix which in general looks like

$$\begin{bmatrix} I & 0 \\ 0 & 0 \end{bmatrix} = X \quad \text{(say)},$$

except that some of the zero matrices may be missing. We know that the rank of this final matrix is the same as $\mathrm{r}(A) = \mathrm{r}$, say; but this rank is clearly the same as the size of the unit matrix in the top left-hand corner, which could be written as I_{r}. This simple matrix is called a *canonical form* for the original matrix A. We have reached it by applying a long sequence of elementary operations, and these are equivalent to multiplying on the left and right by sequences of nonsingular square matrices. We can combine these multiplying matrices together to form single nonsingular square matrices, P and Q say, and then the canonical form matrix X is equal to PAQ. Being nonsingular, P and Q have inverses, and clearly $A = P^{-1}XQ^{-1}$.

We can use the canonical form straight away to study the rank of the product of two matrices. Suppose A is square and nonsingular. Then its canonical form X is simply a unit matrix, and multiplying another matrix by A is the same as multiplying by $P^{-1}IQ^{-1}$. This corresponds to a sequence of elementary operations and does not affect the rank. More generally, if A is $m \times n$ and of rank r, what can we say about the rank of AC, where C is $n \times q$? We have

$$PAQ = \begin{bmatrix} I_r & 0 \\ 0 & 0 \end{bmatrix},$$

so that

$$PAC = \begin{bmatrix} I_r & 0 \\ 0 & 0 \end{bmatrix} Q^{-1}C = \begin{bmatrix} D \\ 0 \end{bmatrix}$$

where D is some $r \times q$ matrix. But the rank of PAC is the same as that of AC, and we can now see that this cannot exceed r, the rank of A. We have shown that $\mathrm{r}(AC) \leq \mathrm{r}(A)$, and in exactly the same way we can show that $\mathrm{r}(AC) \leq \mathrm{r}(C)$.

4.4 Rank and linear dependence

Again let A be an $m \times n$ matrix of rank $r < n$ such that its canonical form is

$$PAQ = \begin{bmatrix} I_r & 0 \\ 0 & 0 \end{bmatrix}$$

and consider the $n \times (n - r)$ matrix

$$D = \begin{bmatrix} 0 \\ I_{n-r} \end{bmatrix}.$$

Then $PAQ \cdot D = 0$, and if we write C for the matrix QD we have found a matrix C of rank $n - r$ which is such that $AC = 0$. This equation represents $n - r$ linear relationships between the columns of A. We have shown that if the rank of a matrix is less than the number of columns, then the columns are linearly dependent. Clearly the same is true of the rows.

We can show rather more than this. The coefficient matrix C is of rank $n - r$ and so it contains a square nonsingular submatrix of size $n - r$. For convenience in writing the formulae, suppose that this occupies the first $n - r$ rows of C so that we can write.

$$C = \begin{bmatrix} C_1 \\ C_2 \end{bmatrix}$$

with C_1 nonsingular. Now if there were a linear relationship between the columns of C, given by $C\mathbf{h} = \mathbf{0}$ say, it would expand to $C_1\mathbf{h} = \mathbf{0}$ and $C_2\mathbf{h} = \mathbf{0}$, and we can deduce from the first of these that the elements of $\mathbf{h}$ must be all zero. Thus the columns of C are linearly independent and the equation $AC = 0$ represents $n - r$ *linearly independent* conditions on the columns of A. We have shown that the n columns of an $m \times n$ matrix of rank r are subject to $n - r$ linearly independent relations or *constraints*, and incidentally that if $r = n$ the columns are independent.

With C partitioned as above, partition A in the same way so that we can write

$$[A_1 \quad A_2] \begin{bmatrix} C_1 \\ C_2 \end{bmatrix} = 0.$$

Then

$$A_1C_1 + A_2C_2 = 0 \quad \text{and} \quad A_1 = -A_2C_2C_1^{-1}.$$

This shows that if we know the r columns of the submatrix A_2 and the coefficient matrix C of the constraints, we can calculate the remaining columns of A as a set of linear combinations.

Examples 4

4.1 Reduce the matrix

$$\begin{bmatrix} 14 & 7 & 21 & 35 \\ 8 & 8 & 8 & 32 \\ 12 & 5 & 19 & 27 \\ 10 & 4 & 16 & 22 \\ 9 & 6 & 12 & 27 \end{bmatrix}$$

to canonical form, finding the matrices P and Q such that

$$PAQ = \begin{bmatrix} I & 0 \\ 0 & 0 \end{bmatrix}.$$

[We can start by using row 1 to remove the elements in the first column. The operations are

$$\begin{aligned} &\text{row } 2 - \tfrac{8}{14}\,(\text{row } 1), \\ &\text{row } 3 - \tfrac{12}{14}\,(\text{row } 1), \\ &\text{row } 4 - \tfrac{10}{14}\,(\text{row } 1), \\ &\text{row } 5 - \tfrac{9}{14}\,(\text{row } 1). \end{aligned}$$

This gives

$$\begin{bmatrix} 14 & 7 & 21 & 35 \\ 0 & 4 & -4 & 12 \\ 0 & -1 & 1 & -3 \\ 0 & -1 & 1 & -3 \\ 0 & 1.5 & -1.5 & 4.5 \end{bmatrix},$$

and applying the same operations to a unit matrix, the left-hand multiplier must be

$$\begin{bmatrix} 1 & 0 & 0 & 0 & 0 \\ -\frac{8}{14} & 1 & 0 & 0 & 0 \\ -\frac{12}{14} & 0 & 1 & 0 & 0 \\ -\frac{10}{14} & 0 & 0 & 1 & 0 \\ -\frac{9}{14} & 0 & 0 & 0 & 1 \end{bmatrix}.$$

Next we remove the elements in row 1 using column 1. This leaves

$$\begin{bmatrix} 14 & 0 & 0 & 0 \\ 0 & 4 & -4 & 12 \\ 0 & -1 & 1 & -3 \\ 0 & -1 & 1 & -3 \\ 0 & 1.5 & -1.5 & 4.5 \end{bmatrix}$$

with a right-hand multiplier of

$$\begin{bmatrix} 1 & -\frac{7}{14} & -\frac{21}{14} & -\frac{35}{14} \\ 0 & 1 & 0 & 0 \\ 0 & 0 & 1 & 0 \\ 0 & 0 & 0 & 1 \end{bmatrix}.$$

Next remove the elements in columns 3 and 4 using column 2. This leaves

$$\begin{bmatrix} 14 & 0 & 0 & 0 \\ 0 & 4 & 0 & 0 \\ 0 & -1 & 0 & 0 \\ 0 & -1 & 0 & 0 \\ 0 & 1.5 & 0 & 0 \end{bmatrix}$$

and changes the right-hand multiplier to

$$\begin{bmatrix} 1 & -\frac{7}{14} & -\frac{28}{14} & -\frac{14}{14} \\ 0 & 1 & 1 & -3 \\ 0 & 0 & 1 & 0 \\ 0 & 0 & 0 & 1 \end{bmatrix}.$$

Now we can remove the elements in rows 3 to 5 using row 2:

$$\begin{bmatrix} 14 & 0 & 0 & 0 \\ 0 & 4 & 0 & 0 \\ 0 & 0 & 0 & 0 \\ 0 & 0 & 0 & 0 \\ 0 & 0 & 0 & 0 \end{bmatrix}$$

making the left-hand multiplier

$$\begin{bmatrix} 1 & 0 & 0 & 0 & 0 \\ -\frac{8}{14} & 1 & 0 & 0 & 0 \\ -\frac{14}{14} & \frac{1}{4} & 1 & 0 & 0 \\ -\frac{12}{14} & \frac{1}{4} & 0 & 1 & 0 \\ -\frac{6}{14} & -\frac{3}{8} & 0 & 0 & 1 \end{bmatrix}.$$

Finally we absorb the diagonal elements into the left-hand multiplier which becomes

$$\begin{bmatrix} \frac{1}{4} & 0 & 0 & 0 & 0 \\ -\frac{2}{14} & \frac{1}{4} & 0 & 0 & 0 \\ -\frac{14}{14} & \frac{1}{4} & 1 & 0 & 0 \\ -\frac{12}{14} & \frac{1}{4} & 0 & 1 & 0 \\ -\frac{6}{14} & -\frac{3}{8} & 0 & 0 & 1 \end{bmatrix}.$$

The original matrix was of rank 2. What are the relationships between the columns?]

4.2 Show that the rows of the matrix in example 3.2 are subject to the linear conditions

$$\boldsymbol{r}_1 - (\boldsymbol{r}_2 + \boldsymbol{r}_3 + \boldsymbol{r}_4) = \mathbf{0}, \quad \boldsymbol{r}_1 - (\boldsymbol{r}_5 + \boldsymbol{r}_6 + \boldsymbol{r}_7 + \boldsymbol{r}_8) = \mathbf{0},$$

so that the matrix is of nullity 2.

5
Simultaneous equations and generalized inverses

5.1 The existence of solutions

We now consider in detail the study of the simultaneous equations

$$A\boldsymbol{x} = \boldsymbol{h}$$

where we continue to take A to be a square $p \times p$ matrix. If A is non-singular, it has an inverse and we have immediately the solution

$$\boldsymbol{x} = A^{-1}\boldsymbol{h}.$$

Moreover, the uniqueness of the inverse tells us that this is the only solution to the equations. There is little more to be said about this situation, and we turn to the more complex case of A singular, of rank $r < p$ say.

First of all, we know that we can find a $(p-r) \times p$ matrix C such that $CA = 0$. If it turns out that $C\boldsymbol{h} \neq \boldsymbol{0}$, the equations involve a contradiction. There is then no vector $\boldsymbol{x}$ for which the equations are true, and we say that they are *inconsistent*. Here too there is little more to be said and in what follows we shall assume that we are dealing with consistent equations, since these are what arise in statistical applications. This means that the right-hand side must satisfy the condition $C\boldsymbol{h} = \boldsymbol{0}$, so that the elements of $\boldsymbol{h}$ are subject to the same constraints as are the rows of A. It may be noted that A and the $p \times (p+1)$ matrix $[A, \boldsymbol{h}]$ then have the same rank.

We can transform A to its canonical form, obtaining

$$X = PAQ = \begin{bmatrix} I_r & 0 \\ 0 & 0 \end{bmatrix}.$$

Then from the original equations $A\boldsymbol{x} = \boldsymbol{h}$ we can derive

$$PAQ \cdot Q^{-1}\boldsymbol{x} = P\boldsymbol{h}$$

and we can write these equations as $X\boldsymbol{y} = \boldsymbol{k}$ with $\boldsymbol{y} = Q^{-1}\boldsymbol{x}$ and $\boldsymbol{k} = P\boldsymbol{h}$. Let us partition $\boldsymbol{y}$ and $\boldsymbol{k}$ into

$$\boldsymbol{y} = \begin{bmatrix} \boldsymbol{y}_1 \\ \boldsymbol{y}_2 \end{bmatrix}, \qquad \boldsymbol{k} = \begin{bmatrix} \boldsymbol{k}_1 \\ \boldsymbol{k}_2 \end{bmatrix},$$

so as to separate off the first r elements of each. Then the equations break down into two parts:

$$Iy_1 = k_1, \qquad 0y_2 = k_2.$$

If the equations are to be consistent, the second of these shows that we must have $k_2 = 0$. But when this is so, the elements of y_2 are completely arbitrary: *any* vector y whose first r elements are the same as those of k_1 will satisfy the equations. If y is any solution of $Xy = k$, then $x = Qy$ is a solution of the original equations $Ax = h$. We have shown that when A is of rank $r<p$ and when the equations are consistent, then there is a $(p-r)$-fold infinity of solutions: we can fix $p-r$ elements of x arbitrarily and then calculate the other elements so as to make x a solution.

5.2 Generalized inverses

From the way we have derived the solutions above, it is clear that the elements of the solution vector x are linear combinations of the elements of the right-hand side vector h. Whether A is singular or not, the solutions to the consistent equations $Ax = h$ can be written in the form $x = Bh$ for some matrix B. If B is a matrix such that Bh satisfies the equations $Ax = h$ whenever these are consistent, we say that B is a *generalized inverse* or *g-inverse* of A and denote it by A^-. If A is nonsingular, then B is simply the ordinary inverse A^{-1} and is unique. If A is singular, the fact that the equations have many solutions shows that the g-inverse is not unique (though we shall show that it has certain unique features). One particular g-inverse (known as the Moore–Penrose inverse) can be singled out by imposing certain extra conditions, but this is not of any great interest in statistical applications.

If A is singular we cannot have $AA^- = I$ since then A^- would be a genuine inverse and we know that there is none. What we do have is the important relation $AA^-A = A$. We can prove this one column at a time. Let a be any column of A; then the elements of a are subject to the same constraints are as the whole rows of A, so that the equations $Ax = a$ are consistent and have the solution $x = A^-a$. But this means that $A(A^-a) = a$, and now, putting all the columns together, we have $AA^-A = A$ as required. Conversely, if B is any matrix such that $ABA = A$, then B is a g-inverse of A. For if $Ax = h$, then $ABAx = Ax = h$, so that $AB(Ax) = h$, $ABh = h$. Writing this as $A(Bh) = h$ shows that Bh is a solution to the equations. This relation provides a useful method of checking whether one matrix is a g-inverse of another.

Now let A be an $m \times n$ matrix of rank r and write its canonical form in the usual way as

$$X = PAQ = \begin{bmatrix} I_r & 0 \\ 0 & 0 \end{bmatrix}$$

It is easily verified that $XX^{\mathsf{T}}X = X$, and so X^{T} is a g-inverse of X by the result in the previous paragraph. We have $A = P^{-1}XQ^{-1}$ and we find that $QX^{\mathsf{T}}P$ (an $n \times m$ matrix) is a g-inverse of A, since

$$\begin{aligned} A(QX^{\mathsf{T}}P)A &= P^{-1}X(Q^{-1}Q)X^{\mathsf{T}}(PP^{-1})XQ^{-1} \\ &= P^{-1}(XX^{\mathsf{T}}X)Q^{-1} \\ &= P^{-1}XQ^{-1} \\ &= A. \end{aligned}$$

The matrix A, being of rank r, contains a nonsingular $r \times r$ submatrix; suppose for convenience that this occurs in the top left-hand corner, so that we can write A in partitioned form as

$$\begin{bmatrix} A_{11} & A_{12} \\ A_{21} & A_{22} \end{bmatrix}$$

with A_{11} $r \times r$ and nonsingular. Then it can be verified that the partitioned matrix

$$\begin{bmatrix} C_{11} & C_{12} \\ C_{21} & C_{22} \end{bmatrix},$$

where the submatrices C_{12}, C_{21}, and C_{22} are entirely arbitrary, is a g-inverse of A provided only that C_{11} is given by

$$C_{11} = A_{11}^{-1} - A_{11}^{-1}A_{12}C_{21} - C_{12}A_{21}A_{11}^{-1} - A_{11}^{-1}A_{12}C_{22}A_{21}A_{11}^{-1}.$$

Various results flow from this. A particularly simple g-inverse is provided by taking C_{12}, C_{21}, and C_{22} all to be zero matrices: we just have to find a nonsingular $r \times r$ submatrix of A, invert it and augment the inverse by a suitable number of rows and columns consisting of zeros. We see also that since C_{11} is nonsingular, the rank of the g-inverse is at least r. By taking C_{12} and C_{21} to be zero matrices and C_{22} to be a diagonal matrix with 0's and 1's on its diagonal, we see that we can produce a g-inverse for A with any rank from r up to the smaller of m and n.

The square matrix $H = A^{-}A$ is of some importance. If A is nonsingular, we have of course that $H = I$; but in any case we find that

$$\begin{aligned} H^2 &= A^{-}AA^{-}A \\ &= A^{-}(AA^{-}A) \\ &= A^{-}A \\ &= H. \end{aligned}$$

A square matrix whose square is equal to itself is called *idempotent*. Because $H = A^-A$ we know that $r(H) \leq r(A)$; but we also have that $A = AH$ so that $r(A) \leq r(H)$ and we deduce that H and A must have the same rank. We also find that

$$\begin{aligned}(I - H)^2 &= I^2 - 2H + H^2\\ &= I - 2H + H\\ &= I - H\end{aligned}$$

so that $I - H$ is also idempotent. It can also be shown that $r(H) + r(I - H) = p$.

Let A be singular and the equations $A\boldsymbol{x} = \boldsymbol{h}$ be consistent, and suppose that one particular solution is given by

$$\boldsymbol{x} = A_0^-\boldsymbol{h}$$

where A_0^- is one particular g-inverse. Then all the other solutions can be written in the form

$$\boldsymbol{x} = A_0^-\boldsymbol{h} + (I - H_0)\boldsymbol{g}$$

where $H_0 = A_0^-A$ and $\boldsymbol{g}$ is a completely arbitrary vector. That this $\boldsymbol{x}$ is a solution follows at once by multiplying it on the left by A. For the converse, suppose that the vector $\boldsymbol{y}$ is a solution, so that $A\boldsymbol{y} = \boldsymbol{h}$. Then $A_0^-\boldsymbol{h} - H_0\boldsymbol{y} = \boldsymbol{0}$ so that we can write

$$\begin{aligned}\boldsymbol{y} &= A_0^-\boldsymbol{h} + \boldsymbol{y} - H_0\boldsymbol{y}\\ &= A_0^-\boldsymbol{h} + (I - H_0)\boldsymbol{y}\end{aligned}$$

which is in the required form.

If A is singular, we know that certain conditions on the vector $\boldsymbol{h}$ are necessary for the equations $A\boldsymbol{x} = \boldsymbol{h}$ to be consistent. We can express these conditions explicitly in terms of g-inverses. It is in fact both necessary and sufficient that $AA^-\boldsymbol{h} = \boldsymbol{h}$ for some g-inverse A^- (and hence, it turns out, for any g-inverse A^-) since if the equations are consistent, they will have a solution $\boldsymbol{x}$. Then

$$\begin{aligned}AA^-\boldsymbol{h} &= AA^-(A\boldsymbol{x})\\ &= (AA^-A)\boldsymbol{x}\\ &= A\boldsymbol{x}\\ &= \boldsymbol{h}\end{aligned}$$

as required. Conversely if the condition holds and $\boldsymbol{x} = A^-\boldsymbol{h}$, then $A\boldsymbol{x} = AA^-\boldsymbol{h} = \boldsymbol{h}$ so that the equations are consistent with $\boldsymbol{x}$ as a solution.

As another approach, because $AA^-A = A$ we have

$$(AA^- - I)A = 0.$$

This means that the matrix $AA^- - I$ is such that its nonzero rows contain the coefficients of the linear constraints to which the rows of A are subject. But the condition for consistency is that the elements of $\boldsymbol{h}$ are subject to the same constraints, i.e. that $(AA^- - I)\boldsymbol{h} = \boldsymbol{0}$, or $AA^-\boldsymbol{h} = \boldsymbol{h}$.

If A is $p \times p$ and symmetric (as is usually the case in statistical applications) with $\mathrm{r}(A) = r$, we can easily provide it with a symmetric g-inverse of rank r. The rows and columns of $\boldsymbol{A}$ are subject to the same set of linear constraints. Consequently there must be a nonsingular submatrix of rank r which contains the same rows as columns. It is hence on the main diagonal and so is itself symmetric. We can invert this, giving a symmetric inverse, and then complete the g-inverse by bordering it with 0's as required.

These g-inverses with zero borders have the property that A is a g-inverse for any one of them. For if

$$A = \begin{bmatrix} A_{11} & A_{12} \\ A^{\mathsf{T}}_{12} & A_{22} \end{bmatrix}$$

with A_{11} non-singular, and

$$A^- = \begin{bmatrix} A_{11}^{-1} & 0 \\ 0 & 0 \end{bmatrix},$$

then it is easy to show that $A^-AA^- = A^-$.

5.3 Unique combinations

Although A^- is not unique when A is singular, it has (not unnaturally) certain unique features. As usual, let us consider the consistent equations $A\boldsymbol{x} = \boldsymbol{h}$ with A $p \times p$ and singular, and take A^-_0 as one particular g-inverse. Then we know that any solution of the equations can be written in the form

$$\boldsymbol{x} = A^-_0\boldsymbol{h} + (I - H_0)\boldsymbol{g}$$

with $H_0 = A^-_0A$ and $\boldsymbol{g}$ an arbitrary vector. Now let us consider a linear combination of the elements of $\boldsymbol{x}$ with coefficients in a vector $\boldsymbol{q}$, the scalar quantity $\boldsymbol{q}^{\mathsf{T}}\boldsymbol{x}$. Suppose that the coefficients are such that $\boldsymbol{q}^{\mathsf{T}}H_0 = \boldsymbol{q}^{\mathsf{T}}$. Then

$$\boldsymbol{q}^{\mathsf{T}}\boldsymbol{x} = \boldsymbol{q}^{\mathsf{T}}A_0\boldsymbol{h} + \boldsymbol{q}^{\mathsf{T}}(I - H_0)\boldsymbol{g}$$

and the second term on the right vanishes. This means that $\boldsymbol{q}^{\mathsf{T}}\boldsymbol{x}$ is the same quantity no matter which of the solutions to the equations we choose. The multiplying vector $\boldsymbol{q}^{\mathsf{T}}$ has *annihilated* the arbitrary part of the general solution to the equations. In a statistical context, the quantity of $\boldsymbol{q}^{\mathsf{T}}\boldsymbol{x}$ is called *estimable*. This is because in a least-squares problem with linearly dependent x's, only estimable quantities are uniquely estimated by the data.

Examples 5

5.1 Verify that the equations

$$\begin{aligned} 5x_1 + 3x_2 + 2x_3 &= 4, \\ 3x_1 + 9x_2 - 6x_3 &= 6, \\ 4x_1 + 2x_2 + 2x_3 &= 3, \end{aligned}$$

are consistent and that

$$x_1 = k, \qquad x_2 = 1 - k, \qquad x_3 = 0.5 - k$$

is a solution for any value of k. Write down a g-inverse of the coefficient matrix and show that the rows are subject to the constraint.

$$15(\text{row } 1) - (\text{row } 2) - 18(\text{row } 3) = 0$$

What is the constraint on the columns?
[The 2×2 matrix

$$\begin{bmatrix} 5 & 3 \\ 3 & 9 \end{bmatrix}$$

is nonsingular and has the inverse

$$\begin{bmatrix} 9 & -3 \\ -3 & 5 \end{bmatrix} /36.$$

Thus a g-inverse of the 3×3 matrix is

$$\begin{bmatrix} 9 & -3 & 0 \\ -3 & 5 & 0 \\ 0 & 0 & 0 \end{bmatrix} /36.$$

Calling the coefficient matrix A and this g-inverse A^-, we find that

$$AA^- = \begin{bmatrix} 36 & 0 & 0 \\ 0 & 36 & 0 \\ 30 & -2 & 0 \end{bmatrix}.$$

Thus

$$I - AA^- = \begin{bmatrix} 0 & 0 & 0 \\ 0 & 0 & 0 \\ 15 & -1 & 18 \end{bmatrix} /18$$

with the coefficients of the constraint in row 3.

For the columns,

$$A^{-}A = \begin{bmatrix} 36 & 0 & 36 \\ 0 & 36 & -36 \\ 0 & 0 & 0 \end{bmatrix} /36, \qquad I - A^{-}A = \begin{bmatrix} 0 & 0 & -1 \\ 0 & 0 & 1 \\ 0 & 0 & 1 \end{bmatrix},$$

and the sum of columns 2 and 3 is equal to column 1.]

5.2 The matrix of the sums of squares and products,

$$A = \begin{bmatrix} 58 & 12 & 80 \\ 12 & 42 & -102 \\ 80 & -102 & 466 \end{bmatrix},$$

has rank 2. Write down a symmetric g-inverse A^{-} of rank 2 and check that A is a g-inverse of A^{-}.

5.3 The singular coefficient matrix for the normal equations of a least-squares problem is given in example 3.2. Show that a linear function $\boldsymbol{c}^{\mathrm{T}}\boldsymbol{x}$ of the solutions $\boldsymbol{x}$ is estimable if $\boldsymbol{c}^{\mathrm{T}}\boldsymbol{k} = 0$ with $\boldsymbol{k} = [0, 1, 1, 1, 0, 0, 0, 0]^{\mathrm{T}}$ (a column contrast) or $\boldsymbol{k} = [0, 0, 0, 0, 1, 1, 1, 1]^{\mathrm{T}}$ (a row contrast).

6
Linear spaces

6.1 Definitions

Many results on matrices (especially those of less than full rank) become almost intuitive if they are expressed in geometric terms. As earlier, we think of a vector $\boldsymbol{x}$ with p elements as corresponding to a point in p dimensions with the elements of $\boldsymbol{x}$ as its rectangular coordinates, or perhaps to the line going from the origin to this point (the term 'vector' originally had this meaning of a directed line). If $\boldsymbol{x}_1$ and $\boldsymbol{x}_2$ are two vectors which do not lie in the same direction, i.e. it is not true that $\boldsymbol{x}_1 = k\boldsymbol{x}_2$ for some nonzero constant k, then the two corresponding lines, or the three corresponding points including the origin, define a plane—a two-dimensional space. What is more, any vector of the form $k_1\boldsymbol{x}_1 + k_2\boldsymbol{x}_2$ lies in the same plane, and conversely any vector which lies in the plane can be expressed in this form. Generalizing this, we define a *linear (vector) space* as a set $\mathcal{V}$ of vectors which are such that, if $\boldsymbol{x}_1$ and $\boldsymbol{x}_2$ are any two members of $\mathcal{V}$, then so is the vector $k_1\boldsymbol{x}_1 + k_2\boldsymbol{x}_2$ for any two scalar quantities k_1 and k_2.

Suppose that $\boldsymbol{x}_1, \boldsymbol{x}_2, \ldots, \boldsymbol{x}_m$ are all vectors with p elements. Then the set of all possible vectors of the form

$$k_1\boldsymbol{x}_1 + k_2\boldsymbol{x}_2 + \cdots + k_m\boldsymbol{x}_m$$

constitutes a linear space, and the $\boldsymbol{x}$ vectors are said to *span* or *generate* it. A set of vectors which span a linear space and which are themselves linearly independent are said to form a *basis* for the space (the notion of a basis is a generalization of that of a set of coordinate axes). Suppose that a particular basis for a linear space contains just m vectors. Then we can show that any other basis must contain the same number of vectors, and this number is called the *dimension* of the space. To prove this, suppose that the first basis forms the columns of an $p \times m$ matrix X and the second those of an $p \times n$ matrix Y with $m, n \leqslant p$. Then each column of Y lies in the space and so can be expressed as a linear combination of the columns of X. It follows that we can write $Y = XK$ for some $m \times n$ matrix K. Suppose if possible that $n > m$. Then because the columns of Y are linearly independent, the rank of Y must be n, whereas the rank of the product on the right cannot exceed the smaller of the two ranks, which is m. We thus have a contradiction. In exactly the same way, by reversing the roles of X and Y we can show that n cannot be less than m, and the result $n = m$ is established. The condition $m \leqslant p$ is automatic, of course, because we cannot have more than p linearly independent vectors with p elements. If $m = p$, the space contains all such vectors.

If A is a $p \times p$ matrix, its columns generate a linear space called the *column space* of A; its dimension is the rank of A. The vectors in this space are simply all those of the form $A\boldsymbol{x}$ for some vector of coefficients $\boldsymbol{x}$. If we say that the equations $A\boldsymbol{x} = \boldsymbol{h}$ are consistent if and only if $\boldsymbol{h}$ is in the column space of A, we are doing no more than saying the same thing in different words.

6.2 An orthogonal basis

Suppose we are given a vector space and a set of vectors which constitute a basis for the space. Then we can construct another basis which consists of a set of mutually orthogonal vectors. There are various ways of doing this; one of the simplest is called the *Gram–Schmidt* procedure. Let us write $\boldsymbol{x}_1, \boldsymbol{x}_2, \ldots, \boldsymbol{x}_m$ for the vectors of the original basis. Now construct the vectors of the new basis $\boldsymbol{y}_1, \boldsymbol{y}_2, \ldots, \boldsymbol{y}_m$ as follows

$$\begin{aligned} \boldsymbol{y}_1 &= \boldsymbol{x}_1, \\ \boldsymbol{y}_2 &= \boldsymbol{x}_2 - (\boldsymbol{x}_2^{\mathrm{T}}\boldsymbol{y}_1/\boldsymbol{y}_1^{\mathrm{T}}\boldsymbol{y}_1) \cdot \boldsymbol{y}_1, \\ \boldsymbol{y}_3 &= \boldsymbol{x}_3 - (\boldsymbol{x}_3^{\mathrm{T}}\boldsymbol{y}_2/\boldsymbol{y}_2^{\mathrm{T}}\boldsymbol{y}_2) \cdot \boldsymbol{y}_2 - (\boldsymbol{x}_3^{\mathrm{T}}\boldsymbol{y}_1/\boldsymbol{y}_1^{\mathrm{T}}\boldsymbol{y}_1) \cdot \boldsymbol{y}_1, \end{aligned}$$

and so on.

Let us look more closely at this. We start by making $\boldsymbol{y}_1$ the same as the first $\boldsymbol{x}$-vector. To form $\boldsymbol{y}_2$ we take another $\boldsymbol{x}$-vector and subtract a scalar multiple of $\boldsymbol{y}_1$; straight multiplication will verify that the multiple is just the right one to make $\boldsymbol{y}_2$ orthogonal to $\boldsymbol{y}_1$, i.e. to make $\boldsymbol{y}_1^{\mathrm{T}}\boldsymbol{y}_2 = 0$. For the third $\boldsymbol{y}$-vector we again take an $\boldsymbol{x}$-vector and subtract scalar multiples of the $\boldsymbol{y}$'s we have so far determined. Again, multiplication will show that the multiples are just those which ensure that $\boldsymbol{y}_1^{\mathrm{T}}\boldsymbol{y}_3$ and $\boldsymbol{y}_2^{\mathrm{T}}\boldsymbol{y}_3$ are both zero. The process can be continued until m $\boldsymbol{y}$-vectors have been determined. They will all be mutually orthogonal and so must be linearly independent; they are linear combinations of the $\boldsymbol{x}$'s and so must lie in the space; and consequently they constitute an *orthogonal basis*.

The Gram–Schmidt procedure has an interesting statistical interpretation. It is an important result of linear model or multiple regression theory that if we have a dependent variate y and if we form the regression of y on a number of predictors $x_1, x_2, \ldots$ and calculate the residuals, then the sum of products of the residuals by any one of the x's is zero; in other words, the residuals are orthogonal to any of the x's (see Section 6.5). Now examination of the Gram–Schmidt formula for $\boldsymbol{y}_2$ will show that the scalar quantity by which we have multiplied $\boldsymbol{y}_1$ before the subtraction is just the regression coefficient (through the origin) of $\boldsymbol{x}_2$ on $\boldsymbol{y}_1$—the sum of products divided by the sum of squares. In fact, $\boldsymbol{y}_2$ is formed by taking the regression of x_2 on y_1 and forming the residuals. To get y_3 we need to take the regression of x_3 on y_1 and y_2, which sounds more complicated; but we know that y_1 and y_2 are orthogonal, so that the calculations are no more than those of simple regression.

It is worth pointing out that with the x's linearly independent the Gram–Schmidt procedure cannot be brought to a premature close by one of the y vectors turning out to have zero length. Suppose for instance that $y_k^{\mathrm{T}}y_k = 0$ with $k \leq m$. Then y_k must be a vector of zeros and the kth step in the procedure amounts to showing that a linear combination of the x's with nonzero coefficients vanishes. This contradicts the assumption of linear independence of the x's.

6.3 Projections

Suppose that $\mathcal{V}$ is a vector space of dimension m and that $\mathcal{W}$ is another space of dimension $n < m$ which is contained within $\mathcal{V}$: by this we mean that any vector which lies in the space $\mathcal{W}$ is also a member of $\mathcal{V}$ but not vice versa. Then if x is any vector which is in $\mathcal{V}$ but not in $\mathcal{W}$, we show that x can be written as the sum of two parts, $x = y + z$, where the first vector y is in the space $\mathcal{W}$ and the second vector z is orthogonal to every vector in $\mathcal{W}$. Moreover, we show that y and z are unique.

Choose an orthogonal basis for $\mathcal{W}$ as the vectors $w_1, w_2, \ldots, w_n$ and form the vector

$$z = x - (w_1^{\mathrm{T}}x/w_1^{\mathrm{T}}w_1)w_1 - \cdots - (w_n^{\mathrm{T}}x/w_n^{\mathrm{T}}w_n)w_n.$$

The analogy with the Gram–Schmidt procedure is close and we can easily show that z is orthogonal to each of the w's. This means that it is also orthogonal to any linear combination of the w's and hence to any vector in the $\mathcal{W}$ space.

If we now form $y = x - z$, we see that y is a linear combination of the w's and consequently lies in the $\mathcal{W}$ space. Then $x = y + z$, and y and z have the properties we require. As to the uniqueness, suppose that also $x = y' + z'$. Then

$$(y - y') + (z - z') = x - x = \mathbf{0}.$$

Multiplying this on the left by $(z - z')^{\mathrm{T}}$ and remembering that both y and y' (which are in $\mathcal{W}$) are orthogonal to both of z and z', we get $(z - z')^{\mathrm{T}}(z - z') = 0$. But this scalar quantity is a sum of squares and can only vanish if $z - z' = 0$. Thus $z = z'$ and immediately $y = y'$.

The vector y is called the *orthogonal projection* of x onto the space $\mathcal{W}$. It is easy to visualize the situation when $\mathcal{V}$ is simply the ordinary three-dimensional space and $\mathcal{W}$ is a two-dimensional plane lying in it. Then x

corresponds to a point **X** lying off the plane and $\boldsymbol{y}$ corresponds to the point found by dropping a perpendicular from **X** onto $\mathcal{W}$. The vector z corresponds to the perpendicular itself, which is at right angles to any line lying in the plane.

If z_1 and z_2 are vectors in $\mathcal{V}$ which are both orthogonal to all the vectors in $\mathcal{W}$, then the same is true for the vector $k_1z_1 + k_2z_2$ for any scalar multipliers k_1 and k_2. It follows that the set of vectors with this property form a linear space, which we denote by $\mathcal{W}^\perp$. It is not difficult to show that its dimension is $m - n$. Take as before $\boldsymbol{w}_1, \boldsymbol{w}_2, \ldots, \boldsymbol{w}_n$ to be an orthogonal basis of $\mathcal{W}$ and let $\boldsymbol{u}_1, \boldsymbol{u}_2, \ldots, \boldsymbol{u}_k$ be an orthogonal basis of $\mathcal{W}^\perp$. Then all these vectors lie in the space $\mathcal{V}$ and are linearly independent, so that $n + k \leqslant m$. Suppose that $n + k < m$; then we can make up an orthogonal basis for $\mathcal{V}$ which contains all the $\boldsymbol{w}$ and $\boldsymbol{u}$ vectors plus at least one more. This extra vector would have to be orthogonal to all the vectors in $\mathcal{W}$ and yet not belong to $\mathcal{W}^\perp$. This is a contradiction and the result is proved. What we have done is to carve up the space $\mathcal{V}$ into two subspaces in such a way that any vector in one of the subspaces is orthogonal to any vector in the other, and that any vector in $\mathcal{V}$ can be written as the sum of two vectors, one from each of the subspaces.

6.4 Projections and idempotent matrices

With the same set-up as in the previous section, let $\boldsymbol{x}$ be a vector in a space $\mathcal{V}$ and $\boldsymbol{y}$ its orthogonal projection onto a subspace $\mathcal{W}$. From the Gram–Schmidt way in which $\boldsymbol{y}$ was constructed, it can be seen that $\boldsymbol{y}$ can be written in the form $P\boldsymbol{x}$, the result of multiplying $\boldsymbol{x}$ by a matrix P, and close examination of the procedure shows that P does not depend on $\boldsymbol{x}$ but only on the $\boldsymbol{w}$ vectors which form the basis for $\mathcal{W}$. But suppose now that we try to project onto $\mathcal{W}$ a vector which already lies in $\mathcal{W}$. The geometrical picture makes it clear that this procedure has no effect: the projection is just the same as the original vector. This means that $P\boldsymbol{y} = \boldsymbol{y}$ for any $\boldsymbol{y}$ in $\mathcal{W}$. It follows that $P^2\boldsymbol{x} = P\boldsymbol{x}$ for any $\boldsymbol{x}$ in $\mathcal{V}$, and hence that the matrix P is such that $P^2 = P$. Such a matrix we have called *idempotent*. We also know that the perpendicular vector z is equal to $\boldsymbol{x} - \boldsymbol{y} = \boldsymbol{x} - P\boldsymbol{x}$, and it follows that $(\boldsymbol{x} - P\boldsymbol{x})^\mathrm{T}P\boldsymbol{x} = z^\mathrm{T}\boldsymbol{y} = 0$. Thus $\boldsymbol{x}^\mathrm{T}(P^\mathrm{T} - I)P\boldsymbol{x} = 0$ for any $\boldsymbol{x}$ in $\mathcal{V}$ and consequently $(P^\mathrm{T} - I)P = 0$. This means that $P^\mathrm{T}P = P$ so that P is symmetric. Notice also that $PP^\mathrm{T}P = P$ so that P is its own g-inverse.

We have shown that the projector matrix P is a symmetric idempotent matrix. Conversely, suppose that P is symmetric and idempotent; then it is a projector matrix onto the space spanned by its own columns. We can show this by writing an arbitrary vector $\boldsymbol{x}$ in the form $P\boldsymbol{x} + (I - P)\boldsymbol{x}$. Clearly the first term $\boldsymbol{y} = P\boldsymbol{x}$ is in the column space of $\boldsymbol{P}$ as required. If z is any vector in this column space, it is equal to $P\boldsymbol{u}$ for some vector $\boldsymbol{u}$. Then

$u^T P^T(I - P)x = u^T P^T x - u^T P^T P x$. But if P is idempotent and symmetric, then $P^T P = PP = P$, so that $(u^T P^T)(I - P)x = 0$ and z and $(I - P)x$ are orthogonal. Thus we have expressed x in the form $y + z$ where y and z have the properties defining the projection process.

If we start with an arbitrary rectangular matrix X, the projector onto its column space is given by $X(X^T X)^- X^T$. $(X^T X)$ is symmetric, so that $(X^T X)^-$ can be taken to be symmetric; then $X(X^T X)^- X^T$ is symmetric. It is also idempotent, since $X^T X$ is a g-inverse of $(X^T X)^-$ and so

$$X(X^T X)^- X^T \cdot X(X^T X)^- X^T = X \cdot (X^T X)^- \cdot X^T.$$

Thus $X(X^T X)^- X^T$ is a projector, and if we multiply an arbitrary vector z by it the result is Xg where $g = (X^T X)^- X^T z$, and this lies in the column space of X.

6.5 Least squares

Let a general vector x in a space $\mathscr{V}$ be written as above in the form $y + z$, where y is the projection of x onto a subspace $\mathscr{W}$ and z lies in $\mathscr{W}^\perp$. The vector z is geometrically the perpendicular from the point **X** corresponding to the vector x onto the subspace $\mathscr{W}$, and with this interpretation it is almost obvious that this is the shortest distance from **X** to $\mathscr{W}$: algebraically, $z^T z \leqslant (x - u)^T(x - u)$ for any vector u in $\mathscr{W}$. We can prove this by noting that

$$\begin{aligned}(x - u)^T(x - u) &= (y + z - u)^T(y + z - u) \\ &= (y - u)^T(y - u) + z^T z + (y - u)^T z + z^T(y - u).\end{aligned}$$

The last two products are between a vector in $\mathscr{W}$ and z, a vector in $\mathscr{W}^\perp$, and consequently they both vanish. Moreover, the product $(y - u)^T(y - u)$ is a sum of squares and so is nonnegative. The result we want now follows.

Consider the set of n equations in p unknowns $X\boldsymbol{\beta} = y$ where X is $n \times p$ and suppose that $n > p$ so that the equations are in general inconsistent and have no solution. Can we find a set of values for $\boldsymbol{\beta}$ such that the values of the elements of $X\boldsymbol{\beta}$ are in some sense near to those of y? Suppose that we project y onto the column space of X, writing $y = y_1 + y_2$ with y_1 in this space and y_2 in the space orthogonal to it. Then for any vector y_1 in the column space of X, this choice makes $y_2^T y_2$ a minimum. But this is the same as $(y - X\boldsymbol{\beta})^T(y - X\boldsymbol{\beta})$, and this quantity is the *residual sum of squares*, the sum of the squares of the differences between the observed vector y and the prediction vector $X\boldsymbol{\beta}$. By taking $\boldsymbol{\beta}$ to be a solution of the consistent equations $y_1 = X\boldsymbol{\beta}$ we have minimized the residual sum of squares. For this reason y_1 is referred to as the least-squares estimate of y, and the solutions to the consistent equations are the least-squares estimates of the elements of $\boldsymbol{\beta}$. It may be noted that since the residuals y_2 are in the orthogonal space, we have the important result $Xy_2 = \mathbf{0}$, or in words, the residuals are orthogonal to the predictors. Then

$$X^{\mathrm{T}}\boldsymbol{y} = X^{\mathrm{T}}\boldsymbol{y}_1 = X^{\mathrm{T}}X\boldsymbol{b}$$

where the vector $\boldsymbol{b}$ is the least-squares estimate of $\boldsymbol{\beta}$, so that $\boldsymbol{b} = (X^{\mathrm{T}}X)^{-}X^{\mathrm{T}}\boldsymbol{y}$, the equations being consistent. These are known as the *normal equations* of least-squares. Written explicitly, the predictions are given by $\boldsymbol{y}_1 = X(X^{\mathrm{T}}X)^{-}X^{\mathrm{T}}\boldsymbol{y}$ where $X(X^{\mathrm{T}}X)^{-}X^{\mathrm{T}}$ is the appropriate projector matrix.

6.6 The null space of a singular matrix

If A is a $p \times p$ singular matrix, there are vectors $\boldsymbol{x}$, not equal to 0, such that $A\boldsymbol{x} = \boldsymbol{0}$. If $\boldsymbol{x}_1$ and $\boldsymbol{x}_2$ are two such vectors, then so is $k_1\boldsymbol{x}_1 + k_2\boldsymbol{x}_2$ so that the whole set of such vectors forms a linear space called the *null space* of A. The vectors in the null space may be regarded as coefficients in linear combinations of the columns of A which are equal to zero, hence the dimension of the null space is equal to the nullity of A.

Let A^{-} be a g-inverse of A. Then the solutions of the equations $A\boldsymbol{x} = \boldsymbol{0}$ all satisfy

$$\boldsymbol{x} = \boldsymbol{0} + (I - A^{-}A)\boldsymbol{g}$$

with $\boldsymbol{g}$ an arbitrary vector. The null space of A is thus the column space of the matrix $I - A^{-}A$. In fact we form the general solution of the consistent equations $A\boldsymbol{x} = \boldsymbol{h}$ by taking any particular solution and adding to it an arbitrary vector lying in the null space of A.

6.7 Transformations to orthogonality

We can look at the Gram–Schmidt procedure of Section 6.2 from a purely algebraic point of view. Write the vectors $\boldsymbol{x}_i$ as the columns of an $n \times p$ matrix X. Then we can use the procedure to calculate matrices $Y(n \times p)$ and $R(p \times p)$ such that $X = YR$ where (a) the columns of Y are orthogonal; and (b) R is upper-triangular. With suitable scaling factors incorporated into R, the columns of Y can be taken to be orthonormal. This result can be used in several statistical contexts, notably in least-squares. As before, suppose we start with the n observational equations $X\boldsymbol{\beta} = \boldsymbol{y}$ so that the least-squares estimates are given by the solutions of the normal equations $X^{\mathrm{T}}X\boldsymbol{b} = X^{\mathrm{T}}\boldsymbol{y}$. These are equivalent to

$$R^{\mathrm{T}}Y^{\mathrm{T}}YR\boldsymbol{b} = R^{\mathrm{T}}Y^{\mathrm{T}}\boldsymbol{y},$$

i.e.

$$R^{\mathrm{T}}R\boldsymbol{b} = R^{\mathrm{T}}Y^{\mathrm{T}}\boldsymbol{y}.$$

and since, with linearly independent $\boldsymbol{x}$'s, R is of rank p, these give

$$R\boldsymbol{b} = Y^{\mathrm{T}}\boldsymbol{y},$$

a set of equations with a triangular coefficient matrix which can easily be solved. Numerically speaking, the resulting algorithms turn out to be more accurate than those which involve forming and solving the normal equations.

The transformation $X = YR$ can be done in several ways which may be preferable to the Gram–Schmidt method in certain circumstances. Notice first that we can define Y and R in an alternative manner so that (a) Y $(n \times n)$ is an orthogonal matrix; and (b) R $(n \times p)$ can be partitioned into an upper triangle $(p \times p)$ on top of a zero matrix $((n - p) \times p)$. Then we can write the transformation in the form $Y^{\mathrm{T}}X = R$ and we need to calculate the orthogonal matrix Y.

First we can exploit the interpretation of multiplication by an orthogonal matrix as being equivalent to a rotation of the coordinate axes. This is quite easy if only two axes are involved. Consider the $p \times p$ orthogonal matrix P which is a unit matrix apart from four of the elements:

$$p_{ii} = \cos\theta = p_{jj}, \qquad p_{ij} = \sin\theta = -p_{ji}.$$

Then

$$\begin{aligned} x_{ik} &= \cos\theta \cdot x_{ik} + \sin\theta \cdot x_{jk}, \\ x_{jk} &= -\sin\theta \cdot x_{ik} + \cos\theta \cdot x_{jk}. \end{aligned}$$

Now choose $\tan\theta = x_{jk}/x_{ik}$; then $x'_{jk} = 0$. If row i starts with $q < p$ zero elements, we can successively send to zero the first $q + 1$ elements of row j by multiplying X on the left by suitable P matrices. These are all orthogonal, so that their combined effect is that of multiplication by an orthogonal matrix. With this method, we can first eliminate the first element of row 2 of X, the first two elements of row 3, then the first three elements of row 4, and so on up to row p; thereafter, all the elements of succeeding rows can be eliminated, leaving the triangular matrix augmented by a zero matrix as required. The elementary rotations from which the transformation is built up are sometimes called *Givens* transformations after the originator of the method.

Another kind of orthogonal transformation is due to Householder and is usually referred to by his name. Let $\boldsymbol{u}$ be a vector of unit length so that $\boldsymbol{u}^{\mathrm{T}}\boldsymbol{u} = 1$. Then $P = I - 2\boldsymbol{u}\boldsymbol{u}^{\mathrm{T}}$ can easily be shown to be a symmetric orthogonal matrix. Given any vector $\boldsymbol{x}$, we can choose $\boldsymbol{u}$ so that the vector $P\boldsymbol{x}$ has all its elements equal to zero except the first—geometrically we rotate so that one of the axes lies along the direction of $\boldsymbol{x}$. Let $\boldsymbol{e}$ be the vector $[1, 0, 0, \ldots, 0]^{\mathrm{T}}$ and suppose that $P\boldsymbol{x} = \mathrm{r}\boldsymbol{e}$. Then $\boldsymbol{x}^{\mathrm{T}}\boldsymbol{x} = \mathrm{r}^2$. Also $\boldsymbol{x} - 2(\boldsymbol{u}^{\mathrm{T}}\boldsymbol{x})\boldsymbol{u} = \mathrm{r}\boldsymbol{e}$, so that if x_i and u_i are the ith elements of $\boldsymbol{x}$ and $\boldsymbol{u}$ we must have $\boldsymbol{u}^{\mathrm{T}}\boldsymbol{x} = -\mathrm{r}u_1$. Then $\boldsymbol{x} + 2\mathrm{r}u_1\boldsymbol{u} = \mathrm{r}\boldsymbol{e}$ so that

$$2u_1^2 = 1 - x_1/\mathrm{r}, \; 2u_i = -x_i/(\mathrm{r}u_1) \quad (i > 1).$$

If we give r the opposite sign to that of x_1, the quantity $2u_1^2$ is guaranteed positive. Now if we start from the matrix X we can apply p successive Householder transformations. The first eliminates all but the first element in column one; the second all but the first two in column two; and so on.

Examples 6

6.1 Suppose that the vectors

$$\boldsymbol{x}_1 = [1\ 1\ 1\ 1\ 1]^{\mathrm{T}}; \quad \boldsymbol{x}_2 = [1\ 2\ 3\ 4\ 5]^{\mathrm{T}}, \quad \boldsymbol{x}_3 = [1\ 4\ 9\ 16\ 25]^{\mathrm{T}}$$

are a basis for a space of three dimensions. Use the Gram–Schmidt procedure to produce an orthogonal basis. Compare the results with example 3.3c.

6.2 If $\boldsymbol{x}$ is any vector and $\mathbf{1}$ is a vector whose elements are all equal to 1, check that $\boldsymbol{x} - \mathbf{1}\mathbf{1}^{\mathrm{T}}\boldsymbol{x}/\mathbf{1}^{\mathrm{T}}\mathbf{1}$ is orthogonal to $\mathbf{1}$. This calculation may be described as 'orthogonalizing $\boldsymbol{x}$ to a constant'. What is it in more familiar statistical terminology?

6.3 From the following data:

x	1	3	4	7
y	12	15	22	38,

calculate the simple regression of y on x and verify that the residuals are orthogonal to the x-values. If the fitted values are given by $\boldsymbol{P}\boldsymbol{y}$, where $\boldsymbol{y}$ contains the y-values, find $\boldsymbol{P}$ and check that it is idempotent. What is the rank of $\boldsymbol{P}$?

6.4 The coefficient matrix in example 3.2 has nullity 2. Find a basis for its null space.

6.5 The data matrix

$$X = \begin{bmatrix} -6.86 & 3.20 & 1.32 \\ 9.04 & 6.50 & 0.72 \\ -1.86 & -10.60 & -2.68 \\ 0.74 & -10.20 & -2.98 \\ 1.64 & -0.90 & -0.78 \\ -6.26 & 6.60 & 3.02 \\ 3.24 & -1.30 & -1.28 \\ 3.24 & 2.70 & 0.32 \\ 0.64 & 7.30 & 1.62 \\ -3.56 & -3.30 & 0.72 \end{bmatrix}$$

contains the data from Fig. 1.1(a) expressed as deviations from column means. Use the Gram–Schmidt procedure to express X in the form $X = QU$ with Q having orthogonal columns and U being upper-triangular. [First orthogonalize columns 2 and 3 to column 1. The multipliers are 27.14/208.74 and −24.59/208.74, and we get

$$X = \begin{bmatrix} -6.86 & 4.09 & 0.51 \\ 9.04 & 6.50 & 1.78 \\ -1.86 & -10.36 & -2.90 \\ 0.74 & -10.30 & -2.89 \\ 1.64 & -1.11 & -0.59 \\ -6.26 & 7.41 & 2.28 \\ 3.24 & -1.72 & -0.90 \\ 3.24 & 2.28 & 0.70 \\ 0.64 & 7.22 & 1.70 \\ -3.56 & -2.84 & 0.30 \end{bmatrix} \begin{bmatrix} 1 & 0.1300 & -0.1178 \\ 0 & 1 & 0 \\ 0 & 0 & 1 \end{bmatrix}.$$

Next orthogonalize column 3 to column 2. The multiplier is 103.48/382.94 and we get

$$X = \begin{bmatrix} -6.86 & 4.09 & -0.60 \\ 9.04 & 5.32 & 0.34 \\ -1.86 & -10.36 & -0.10 \\ 0.74 & -10.30 & -0.11 \\ 1.64 & -1.11 & -0.29 \\ -6.26 & 7.41 & 0.28 \\ 3.24 & -1.72 & -0.44 \\ 3.24 & 2.28 & 0.08 \\ 0.64 & 7.22 & -0.25 \\ -3.56 & -2.84 & 1.07 \end{bmatrix} \begin{bmatrix} 1 & 0.1300 & -0.1178 \\ 0 & 1 & 0.2702 \\ 0 & 0 & 1 \end{bmatrix}.$$

Note that the last column now contains the residuals from the regression of arm circumference on height and weight.

If required we can standardize the orthogonal columns to unit length, absorbing the necessary factors into the triangular matrix:

$$X = \begin{bmatrix} -0.4748 & 0.2090 & -0.4173 \\ 0.6257 & 0.2719 & 0.2365 \\ -0.1287 & -0.5294 & -0.0695 \\ 0.0512 & -0.5263 & -0.0765 \\ 0.1135 & -0.0567 & -0.2017 \\ -0.4333 & 0.3787 & 0.1947 \\ 0.2243 & -0.0879 & -0.3060 \\ 0.2243 & 0.1165 & 0.0556 \\ 0.0443 & 0.3690 & -0.1739 \\ -0.2464 & -0.1451 & 0.7441 \end{bmatrix} \begin{bmatrix} 14.45 & 1.88 & -1.70 \\ 0 & 19.57 & 5.29 \\ 0 & 0 & 1.44 \end{bmatrix}.]$$

7
Quadratic forms and eigensystems

7.1 Quadratic forms

We have so far concentrated on linear problems, mostly associated with the simultaneous equations that arise in least-squares (multiple regression) methodology. Now we turn to quadratic functions of the data which are of importance in the theory of the analysis of variance and which also occur in most techniques of multivariate analysis.

The scalar quantity $\boldsymbol{x}^{\mathrm{T}}A\boldsymbol{x}$ is called a *quadratic form* in the variables $\boldsymbol{x}$. The matrix A is square and we may as well take it to be symmetric, since the term in $\boldsymbol{x}_i\boldsymbol{x}_j$ has coefficient $a_{ij} + a_{ji}$, and there is no reason why we should not take a_{ij} and a_{ji} to be equal. Typically in statistical applications A will be a matrix of variances and covariances or of sums of squares and products.

A particularly simple quadratic form which we have met already occurs when $A = I$. This is just $\boldsymbol{x}^{\mathrm{T}}\boldsymbol{x}$, the squared length of the vector $\boldsymbol{x}$. This is positive for all vectors other than $\boldsymbol{x} = \boldsymbol{0}$, and such a form is called *positive definite*. If the form is never negative but is equal to 0 for some $\boldsymbol{x} \neq \boldsymbol{0}$, we say that it is *positive semidefinite* or *nonnegative definite*. We use the same terminology for a quadratic form and its coefficient matrix; thus we talk of a positive definite matrix and of the rank of a form.

Statistically, suppose that A is the matrix of variances and covariances of a set of variables $x_1, x_2, \ldots, x_p$. Then the variance of a linear combination of the x's, say $\boldsymbol{c}^{\mathrm{T}}\boldsymbol{x}$, is given by $\boldsymbol{c}^{\mathrm{T}}A\boldsymbol{c}$. This cannot be negative and can only be zero if the x's are linearly dependent, with an exact linear relationship between them. It follows that A must be at least positive semidefinite and will usually be positive definite.

7.2 A square-root matrix

If A is positive definite of size $p \times p$, we can find an upper triangular matrix U such that $A = U^{\mathrm{T}}U$, so that U is a sort of square root of A. This can be seen most easily by giving the rule for the construction of U, which is simply

$$(\text{column } i \text{ of } U) \times (\text{column } j \text{ of } U) = a_{ij},$$

Using this, we can calculate the elements of U from the following set of relations:

$$\begin{aligned} u_{11}^2 &= a_{11}, \\ u_{11}u_{12} &= a_{12}, \\ u_{11}u_{13} &= a_{13}, \\ \ldots \\ u_{12}^2 + u_{22}^2 &= a_{22}, \\ u_{12}u_{13} + u_{22}u_{23} &= a_{23}, \\ \ldots \\ u_{13}^2 + u_{23}^2 + u_{33}^2 &= a_{33}, \end{aligned}$$

and so on. In each of these equations all but one of the u's has been calculated in the equations above.

When A is a matrix of sums of squares and products of a set of variables $x_1, x_2, \ldots, x_p$ (as in a multiple regression problem), the elements of U have a statistical interpretation. It will be seen that u_{ii}^2 is the residual sum of squares of x_i after allowing for the regression of x_i on $x_{i-1}, x_{i-2}, \ldots, x_1$. In the same way, $u_{ii}u_{ij}$ is the residual sum of products of x_i and x_j after both of them have been regressed on the same set of previous variables. With A positive definite, there is no exact linear relationship between the x's, and $|U| = \sqrt{|A|} \neq 0$. Consequently the procedure cannot fail through one of the u_{ii}^2's becoming zero or negative.

A statistical application of the square-root matrix arises if we again take A to be a matrix of variances and covariances of a set of variables $\boldsymbol{x}$ (which we assume to be linearly independent) and transform to a new set of variables given by $\boldsymbol{y} = (U^{\mathrm{T}})^{-1}\boldsymbol{x}$. The dispersion matrix of these new variables is found to be

$$(U^{\mathrm{T}})^{-1} \cdot U^{\mathrm{T}}U \cdot U^{-1} = I$$

so that the y's have equal variances and zero covariances. This transformation can be used to simplify many multivariate distribution problems.

The procedure for forming U (called the square-root or *Cholesky* procedure) is also of interest as the basis for an excellent numerical method for solving simultaneous linear equations and inverting matrices (see Chapter 8). In fact, if $V^{\mathrm{T}} = U^{-1}$, then V is also triangular, and $A^{-1} = V^{\mathrm{T}}V$. The algorithms for forming U and V are both simple and numerically precise.

The square-root procedure can also be applied when A is only positive semidefinite. The determinant of U is $\Pi_i u_{ii}$ and $|U|^2 = |A|$, so that if A is singular then so is U and one or more of the u_{ii} must be zero. This means that the algorithm as presented will fail with a division by zero. However, if when a zero u_{ii} occurs we simply set all subsequent elements in the same row of U to zero, the relation $A = U^{\mathrm{T}}U$ will still hold and the matrix $U^{-}(U^{\mathrm{T}})^{-}$ will be a g-inverse of A. In addition, the variates $\boldsymbol{y} = (U^{\mathrm{T}})^{-}\boldsymbol{x}$ will still be uncorrelated and will have equal variances apart from those whose variances are zero.

7.3 The characteristic equation

Suppose that A is $p \times p$ and that we transform a vector $\boldsymbol{x}$ to $\boldsymbol{x} = A\boldsymbol{x}$. Generally speaking, this will change both the length and the direction of the vector, but it is possible that for some vectors $\boldsymbol{x}$ the result is merely an expansion or contraction so that $\boldsymbol{y}$ is a scalar multiple of $\boldsymbol{x}$, say $\boldsymbol{y} = \lambda\boldsymbol{x}$. If this is to be so, we must have $A\boldsymbol{x} = \lambda\boldsymbol{x}$ or $(A - \lambda I)\boldsymbol{x} = \mathbf{0}$. If the matrix $A - \lambda I$ is nonsingular, the unique solution to these equations is $\boldsymbol{x} = \mathbf{0}$, so that we only get a nontrivial solution when $|A - \lambda I| = 0$. If we expand this determinant, the condition becomes a polynomial equation in λ of degree p. This is called the *characteristic equation* of A. Its p roots (which may be real or complex, simple or multiple) are called the *eigenvalues* (or proper values, or characteristic values, or latent roots) of A. If λ is an eigenvalue, a nonzero vector $\boldsymbol{x}$ satisfying $A\boldsymbol{x} = \lambda\boldsymbol{x}$ is called an *eigenvector* (proper vector, characteristic vector, latent vector) corresponding to λ. It should be noted that $\boldsymbol{x}$ is only determined up to an arbitrary scalar multiple, since if $\boldsymbol{x}$ is an eigenvector corresponding to an eigenvalue λ, then clearly so is $k\boldsymbol{x}$ for any nonzero k. It is often convenient to normalize each eigenvector to have a squared length of 1.

We shall often need to talk about a vector $\boldsymbol{x}$ which is an eigenvector corresponding to an eigenvalue λ. With apologies to the German language, we shall express this by saying that $\boldsymbol{x}$ and λ are eigen to each other.

If we consider the individual terms in the expansion of the determinant $|\boldsymbol{A} - \lambda \boldsymbol{I}|$ we see that the only one containing λ^p is that coming from the product of the diagonal elements of the matrix, and the coefficient of this power in the equation is $(-1)^p$. Similarly, any term containing λ^{p-1} must contain $p - 1$ of the diagonal elements and so must contain all of them. The $p - 1$ elements which contribute λ's can be chosen in p different ways and so the coefficient of λ^{p-1} in the equation must be

$$(-1)^{p-1}(a_{11} + a_{22} + \cdots + a_{pp}) = (-1)^{p-1}\operatorname{tr} A.$$

The other coefficients in the equation can be worked out in the same way, but the most interesting one is the constant term, which is simply $|A|$. From the theory of equations, we can now deduce that the trace of the matrix is equal to the sum of the eigenvalues and the determinant is equal to their product. This shows at once that a singular matrix must have at least one zero eigenvalue. Indeed, if A is singular, we know that there is a nonzero matrix C such that $AC = 0$, and then the nonzero columns of C must all be eigen to eigenvalues of zero. Conversely, a matrix with one or more zero eigenvalues must have a zero determinant and be singular.

7.4 Symmetric matrices

The theory of eigenvalues and vectors is a good deal simpler if the matrix A that we are concerned with is symmetric. This will be true for almost all applications in linear modelling and multivariate analysis where A will be a dispersion or correlation matrix, and for the time being we shall assume it to be true. For a start, all the eigenvalues are real and the elements of the eigenvectors can be taken to be real also. To prove this, suppose that

$$A(\boldsymbol{x} + \mathrm{i}\boldsymbol{y}) = (\lambda + \mathrm{i}\mu)(\boldsymbol{x} + \mathrm{i}\boldsymbol{y})$$

with λ, μ, $\boldsymbol{x}$, and $\boldsymbol{y}$ all real and with $\boldsymbol{x}$ and $\boldsymbol{y}$ not both zero. Then

$$A\boldsymbol{x} = \lambda\boldsymbol{x} - \mu\boldsymbol{y}, \qquad A\boldsymbol{y} = \mu\boldsymbol{x} + \lambda\boldsymbol{y}.$$

Hence

$$\begin{aligned} \boldsymbol{y}^{\mathrm{T}}A\boldsymbol{x} &= \lambda\boldsymbol{y}^{\mathrm{T}}\boldsymbol{x} - \mu\boldsymbol{y}^{\mathrm{T}}\boldsymbol{y} \\ &= \boldsymbol{x}^{\mathrm{T}}A\boldsymbol{y} = \mu\boldsymbol{x}^{\mathrm{T}}\boldsymbol{x} + \lambda\boldsymbol{x}^{\mathrm{T}}\boldsymbol{y}. \end{aligned}$$

It follows that $\mu(\boldsymbol{x}^{\mathrm{T}}\boldsymbol{x} + \boldsymbol{y}^{\mathrm{T}}\boldsymbol{y}) = 0$. But the quantity in parentheses must be positive, so that $\mu = 0$ and λ and $\boldsymbol{x}$, both real, are eigen to each other.

If $\boldsymbol{x}$ is eigen to an eigenvalue λ of A, then $\boldsymbol{x}^{\mathrm{T}}A\boldsymbol{x} = \lambda\boldsymbol{x}^{\mathrm{T}}\boldsymbol{x}$. Then $\lambda = \boldsymbol{x}^{\mathrm{T}}A\boldsymbol{x}/\boldsymbol{x}^{\mathrm{T}}\boldsymbol{x}$ with $\boldsymbol{x}^{\mathrm{T}}\boldsymbol{x} > 0$. and so if A is positive definite (or semidefinite) we can deduce that any eigenvalue λ is greater than (or greater than or equal to) zero.

If $\boldsymbol{x}_i$ and $\boldsymbol{x}_j$ are eigen to two unequal eigenvalues λ_i and λ_j, then they are orthogonal. For:

$$\begin{aligned} \boldsymbol{x}_j^{\mathrm{T}}A\boldsymbol{x}_i &= \lambda_i\boldsymbol{x}_j^{\mathrm{T}}\boldsymbol{x}_i \\ &= \boldsymbol{x}_i^{\mathrm{T}}A\boldsymbol{x}_j = \lambda_j\boldsymbol{x}_i^{\mathrm{T}}\boldsymbol{x}_j; \end{aligned}$$

but $\lambda_i \neq \lambda_j$, and so $\boldsymbol{x}_i^{\mathrm{T}}\boldsymbol{x}_j = 0$.

If $\boldsymbol{x}$ is eigen to eigenvalue λ of A, then

$$A \cdot A\boldsymbol{x} = A \cdot \lambda\boldsymbol{x} = \lambda \cdot \lambda\boldsymbol{x}$$

so that $A^2\boldsymbol{x} = \lambda^2\boldsymbol{x}$. In general for $k > 1$ we find that $\boldsymbol{x}$ is an eigenvector of A^k corresponding to an eigenvalue λ^k. This also works for negative powers if these exist; for if $A\boldsymbol{x} = \lambda\boldsymbol{x}$ with A nonsingular and consequently $\lambda \neq 0$, then $\lambda^{-1}\boldsymbol{x} = A^{-1}\boldsymbol{x}$.

Continuing with the $p \times p$ symmetric matrix A, suppose now that $\boldsymbol{z}$ is any nonzero vector. Then the vectors $\boldsymbol{z}, A\boldsymbol{z}, A^2\boldsymbol{x}, \ldots$ span a linear space, and we can show that this space must contain an eigenvector of A. The dimension of the space cannot exceed p, and this means that there must be a nontrivial linear relationship

$$b_0\boldsymbol{z} + b_1A\boldsymbol{z} + \cdots + b_kA^k\boldsymbol{z} = \mathbf{0}$$

with $1 \leqslant k \leqslant p$. Choose k to be the lowest value for which this is true; then $b_k \neq 0$ and we may as well take $b_k = 1$. We can write this relationship as

$$s(A) \cdot z = \mathbf{0}$$

where $s(x)$ is an ordinary polynomial of degree k. This polynomial can be factorized in the usual way and we can write $s(x) = (x - \lambda)t(x)$, where λ is a root of the equation $s(x) = 0$ and $t(x)$ is a polynomial of degree $k - 1$. Now put

$$\boldsymbol{y} = t(A) \cdot z.$$

Then $\boldsymbol{y} \neq \mathbf{0}$, since otherwise we have a non-trivial linear relationship not involving $A^k z$; and

$$(A - \lambda)\boldsymbol{y} = (A - \lambda)t(A) \cdot z = s(A) \cdot z = \mathbf{0}.$$

Hence $\boldsymbol{y}$ is an eigenvector as required. Its eigenvalue λ is real, since A is symmetric, and so y has real elements.

Next, fix one eigenvector $\boldsymbol{x}_1$ and choose any nonzero vector z which is orthogonal to it. Then $A^r\boldsymbol{x}_1 = \lambda^r\boldsymbol{x}_1$, so that $z^T A^r \boldsymbol{x}_1 = 0$ and the whole of the space which is generated by $z, Az, A^2z, \ldots$ is orthogonal to $\boldsymbol{x}_1$. We can show as before that this space must contain an eigenvector, and this eigenvector must be orthogonal to $\boldsymbol{x}_1$—note that we have now not had to assume that the corresponding eigenvalues are unequal. Proceeding in this way, we can show that A must have p mutually orthogonal eigenvectors. If we standardize each one of these to have unit length, we can arrange them as the columns of an orthogonal matrix X. We now can write

$$AX = X\lambda^{\mathrm{d}}$$

where λ^{d} is the diagonal matrix with the eigenvalues of A on its diagonal. Now $X^{-1} = X^{\mathrm{T}}$, and it follows that A can be written as $X\lambda^{\mathrm{d}}X^{\mathrm{T}}$. Writing this out in full we have

$$A = \lambda_1 \boldsymbol{x}_1 \boldsymbol{x}_1^{\mathrm{T}} + \lambda \boldsymbol{x}_2 \boldsymbol{x}_2^{\mathrm{T}} + \cdots + \lambda_p \boldsymbol{x}_p \boldsymbol{x}_p^{\mathrm{T}}.$$

This is called the *spectral decomposition* of the matrix A. It exhibits the matrix as the sum of p matrices of rank 1, each multiplied by a scalar factor (which may of course be zero). In some expositions of statistical factor analysis, each of these matrices is called a *hierarchy*.

An eigenvalue may be a k-fold multiple root of the charactristic equation. It is then called an eigenvalue of *multiplicity* k, and there will be k mutually orthogonal eigenvectors associated with it. But it can be seen that any linear combination of these eigenvectors is also eigen to the same eigenvalue, so that it is more appropriate to say that the multiple root is associated with a whole linear space of eigenvectors of dimension k, the basis of the space not being uniquely defined. This holds in particular when the multiple eigenvalue is zero, and the space is then just the null space of the matrix.

When A is nonsingular, the spectral decomposition of A^{-1} is

$$\lambda_1^{-1}\boldsymbol{x}_1\boldsymbol{x}_1^{\mathrm{T}} + \lambda_2^{-1}\boldsymbol{x}_2\boldsymbol{x}_2^{\mathrm{T}} + \cdots + \lambda_p^{-1}\boldsymbol{x}_p\boldsymbol{x}_p^{\mathrm{T}}.$$

It is remarkable that when A is singular a g-inverse of A can be obtained from its spectral decomposition in exactly the same way, simply omitting the terms for which $\lambda_i = 0$.

7.5 Orthogonal reduction to diagonal form

Let A be a $p \times p$ symmetric matrix and let P be $p \times p$ and orthogonal. Consider the matrix $B = P^{\mathrm{T}}AP$. The inverse of P is equal to its transpose, so $A = PBP^{\mathrm{T}}$ and if $A\boldsymbol{x} = \lambda\boldsymbol{x}$, then $PB(P^{\mathrm{T}}\boldsymbol{x}) = \lambda\boldsymbol{x}$, $B \cdot P^{\mathrm{T}}\boldsymbol{x} = \lambda P^{\mathrm{T}}\boldsymbol{x}$, and $P^{\mathrm{T}}\boldsymbol{x}$ is an eigenvector of B corresponding to an eigenvalue λ. We say that B is an *orthogonal transformation* of A. We have shown that the process does not affect the eigenvalues, while the eigenvectors change in a fairly straightforward way.

A special orthogonal transformation is implicit in the work of the previous section. Suppose that for the orthogonal matrix we use an X whose columns are the standardized eigenvectors of A. Then we have $X^{\mathrm{T}}AX = \lambda^{\mathrm{d}}$, the diagonal matrix containing the eigenvalues. The results of this orthogonal transformation is thus a diagonal matrix. If A is the dispersion matrix of a set of variates $\boldsymbol{x}$ and $\boldsymbol{y} = P^{\mathrm{T}}\boldsymbol{x}$, then $B = P^{\mathrm{T}}AP$ is the dispersion matrix of the $\boldsymbol{y}$ variates. With $P = X$, this dispersion matrix is diagonal, so that the $\boldsymbol{y}$ variates are uncorrelated. As we shall see in the next section, this transformation has further properties which can be interpreted statistically.

Since X is orthogonal, it and X^{T} are nonsingular and so A and λ^{d} have the same rank. It follows that a symmetric matrix of nullity n has just n zero eigenvalues. This agrees with the fact that the eigenvectors eigen to these eigenvalues form the null space of the matrix. Furthermore, if $\boldsymbol{z}$ is any vector, then $\boldsymbol{z}^{\mathrm{T}}A\boldsymbol{z} = \boldsymbol{w}^{\mathrm{T}}\lambda^{\mathrm{d}}\boldsymbol{w}$ where $\boldsymbol{w} = X^{\mathrm{T}}\boldsymbol{z}$. This in turn is equal to $\Sigma\lambda_i w_i^2$, and it follows that a matrix is positive definite (or semidefinite) if and only if all its eigenvalues are positive (or nonnegative).

7.6 Extremal properties

The standardized eigenvectors $\boldsymbol{x}_1, \boldsymbol{x}_2, \ldots, \boldsymbol{x}_p$ of a $p \times p$ symmetric matrix A are mutually orthogonal and span the whole p-dimensional space. Then any vector $\boldsymbol{z}$ can be written as a linear combination of the eigenvectors, $\boldsymbol{z} = \Sigma c_i\boldsymbol{x}_i$ say, and $A\boldsymbol{z} = \Sigma c_i\lambda_i\boldsymbol{x}_i$ where the λ_i are the corresponding eigenvalues. Now consider the two quadratic forms $\boldsymbol{z}^{\mathrm{T}}A\boldsymbol{z}$ and $\boldsymbol{z}^{\mathrm{T}}\boldsymbol{z}$. They are equal respectively to $\Sigma c_i^2\lambda_i$ and Σc_i^2 and their ratio is $\Sigma c_i^2\lambda_i/\Sigma c_i^2$ which is a weighted mean of the λ_i's with nonnegative weights. The maximum of this ratio for different vectors z occurs when the weight of the largest eigenvalue is 1 with all the other weights zero, and the minimum occurs in a similar way. Thus the vectors which maximize and minimize the ratio are simply those which are eigen to the largest and smallest eigenvalues respectively, and the maximum and minimum values of the ratio are the eigenvalues themselves. Put another way, these eigenvectors maximize and minimize the quadratic form z^TAz subject to the condition $z^Tz = 1$.

Going a step further, let us write λ_1 for the largest eigenvalue and λ_2 for the next largest, with x_1 and x_2 as their eigenvectors. Then, using the spectral decomposition, λ_2 is the largest eigenvalue of the matrix $A - \lambda_1 x_1 x_1^T$ and so x_2 is the vector which maximizes the quadratic form $z^T(A - \lambda_1 x_1 x_1^T)z = z^TAz - \lambda_1 z^T x_1 \cdot x_1^T z$. This means that x_2 is the vector z which maximizes the ratio z^TAz/z^Tz subject to the condition that $z^Tx_1 = 0$, i.e. that z and x_1 are orthogonal. The remaining eigenvalues and vectors can be interpreted in a similar way.

These results have a well-known geometrical interpretation. The equation $x^TAx = c$, a constant, is that of a *quadric*, a p-dimensional object which is a generalization of a conic section (ellipse, parabola, or hyperbola) and whose centre is at the origin. The *principal axes* of the quadric are those directions for which the value of the distance from the centre to a point on the quadric has a turning point. To find these directions we must maximize x^Tx subject to $x^TAx = c$, and introducing μ as a Lagrange multiplier we are led to differentiate $x^Tx - \mu(x^TAx - c)$. Equating the differentials to zero leads to the equations $(I - \mu A)x = \mathbf{0}$ or $(A^{-1} - \mu I)x = \mathbf{0}$ if A is nonsingular. The directions of the principal axes are thus given by the eigenvectors of A^{-1}, which are the same as those of A. Suppose that x_i is a standardized eigenvector corresponding to a nonzero eigenvalue μ_i of A^{-1} and that the point kx_i is on the quadric so that $(kx_i)^TA(kx_i) = c$. Then $kx^T_i \cdot k\mu^{-1}x_i = c$ so that $k^2 = c\mu_i$. It follows that the lengths of the principal axes are all real if the μ_i are all positive, i.e. if A^{-1} (and hence A) is positive definite. The quadric is then an *ellipsoid*. If the smallest eigenvalue of A is λ with corresponding eigenvector x, the longest principal axis (the *major axis*) lies in the direction given by x and its semilength is $c\lambda^{-\frac{1}{2}}$.

Statistically, we can take A to be the dispersion matrix (perhaps normalized to be the correlation matrix) of a set of p variates. The linear functions of the variates with coefficients given by the eigenvectors are the *principal components* of the data. The dispersion matrix of the principal components is simply λ^d. Thus the principal component corresponding to the largest eigenvalue has maximum variance for any linear combination of the x's subject to the sum of squares of the coefficients being equal to 1; the component corresponding to the next largest eigenvalue has maximum variance subject to the same condition and to being uncorrelated with the first; and so on.

7.7 Simultaneous reduction of two symmetric matrices

Suppose we decide to measure distance in a more general metric than the usual one, defining the squared generalized length of a vector $\boldsymbol{x}$ to be the value of a positive definite quadratic form $\boldsymbol{x}^{\mathrm{T}}M\boldsymbol{x}$. Then we can pose the problem of maximizing the value of the form $\boldsymbol{x}^{\mathrm{T}}A\boldsymbol{x}$ subject to the condition $\boldsymbol{x}^{\mathrm{T}}M\boldsymbol{x}$ = constant. This leads to the *generalized eigenvalue problem* of finding vectors $\boldsymbol{x}$ which satisfy the equation $(A - \lambda M)\boldsymbol{x} = \boldsymbol{0}$. We can find a nonsingular matrix U such that $M = U^{\mathrm{T}}U$. Then if $V = U^{-1}$ we have $V^{\mathrm{T}}AV \cdot U\boldsymbol{x} = \lambda U\boldsymbol{x}$. It follows that $y = U\boldsymbol{x}$ is an ordinary eigenvector of the symmetric matrix $B = V^{\mathrm{T}}AV$ with corresponding eigenvalue λ, and $\boldsymbol{x}$ can easily be obtained by calculating $V\boldsymbol{y}$. We know that if $\boldsymbol{y}_1$ and $\boldsymbol{y}_2$ are two eigenvectors of B then $\boldsymbol{y}_1^{\mathrm{T}}\boldsymbol{y}_2 = 0$. Hence if $\boldsymbol{x}_1$ and $\boldsymbol{x}_2$ are the corresponding generalized eigenvectors we have $\boldsymbol{x}_1^{\mathrm{T}}U^{\mathrm{T}}U\boldsymbol{x}_2 = \boldsymbol{x}_1^{\mathrm{T}}M\boldsymbol{x}_2 = 0$. The vectors $\boldsymbol{x}_1$ and $\boldsymbol{x}_2$ are 'orthogonal' in a generalized sense—they may be called M-orthogonal. Furthermore, if we standardize the $\boldsymbol{y}$'s to have unit length with $\boldsymbol{y}^{\mathrm{T}}\boldsymbol{y} = 1$, it follows automatically that the $\boldsymbol{x}$'s are standardized to $\boldsymbol{x}^{\mathrm{T}}M\boldsymbol{x} = 1$, that is to unit generalized length. If we put the standardized generalized eigenvectors together as the columns of a matrix X, we find that simultaneously

$$X^{\mathrm{T}}MX = I \quad \text{and} \quad X^{\mathrm{T}}AX = \boldsymbol{\lambda}^{\mathrm{d}}$$

with $\boldsymbol{\lambda}$ the vector of eigenvalues. We have thus reduced one matrix to the unit matrix and the other to a diagonal matrix by the same M-orthogonal transformation. We can obtain similar results to these when M is only positive semidefinite, since we can still define the square-root matrix U. However, if the rank of M is equal to r, we land up with

$$X^{\mathrm{T}}MX = \begin{bmatrix} I_r & 0 \\ 0 & 0 \end{bmatrix}$$

since the last $p - r$ eigenvalues must be zero.

These results are the basis for *canonical analysis* in its different varieties. We may for example consider multivariate data classified into groups and take A and M to be the between-group and within-group matrices of mean squares and products. Then the problem of finding the linear combination of the variates which has maximum variance subject to the within-group variance being equal to 1 becomes that of solving the equations $(A - \lambda M)\boldsymbol{x} = \boldsymbol{0}$. The canonical variates are formed by taking the elements of the generalized eigenvectors as coefficients. Their within-group dispersion matrix is I and their between-group dispersion matrix is $\boldsymbol{\lambda}^{\mathrm{d}}$, so that they are uncorrelated both within and between groups.

The geometry of this analysis is quite straightforward. We start with a picture of the groups represented by ellipsoidal probability contours (Fig. 7.1). The assumption of a common within-group dispersion matrix

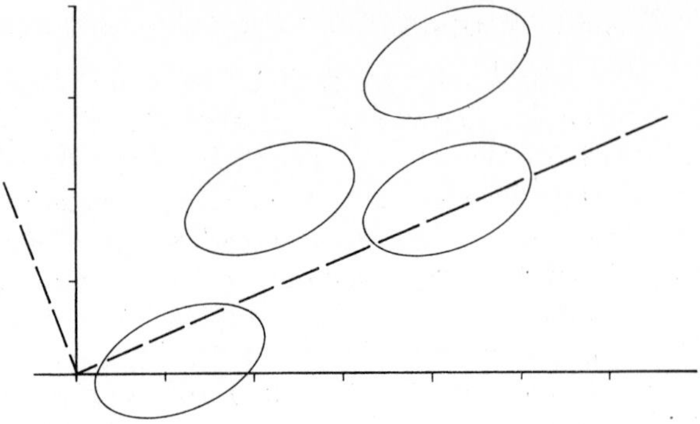

(a)

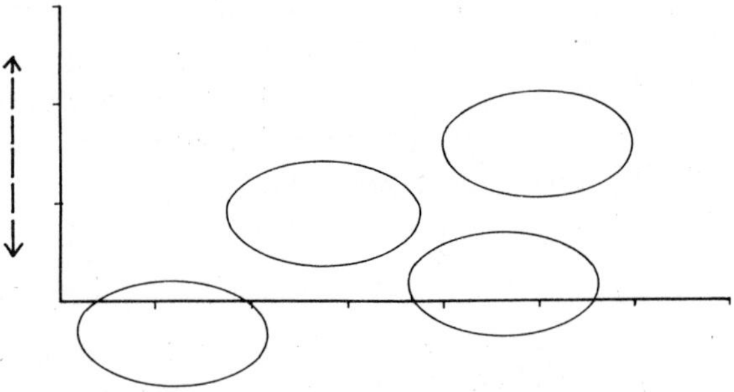

(b)

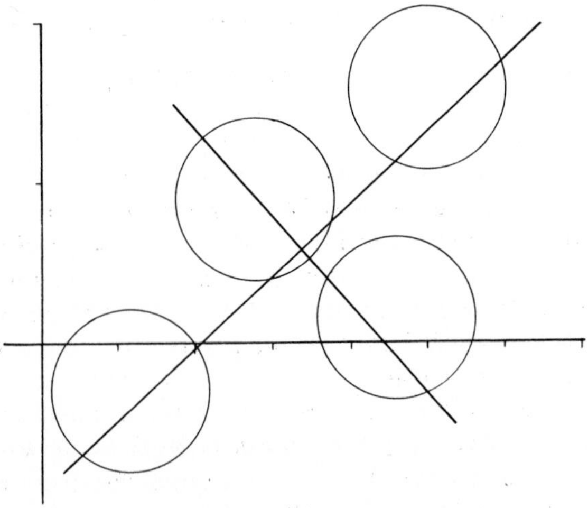

(c)

Fig. 7.1 Canonical analysis. Original axes (a) are related to zero correlation (b) and stretched to equal variances (c) before constructing principal axes.

leads to the ellipsoids being similar in size and orientation. We first make a rigid rotation so that the principal axes of the ellipsoids are parallel to the coordinate axes. Next we rescale each axis separately so that the ellipsoids become spheres. Finally we make a further rotation to align the new axes with the principal axes of the group centres.

7.8 Nonsymmetric matrices

The theory of eigensystems is distinctly more complicated for non-symmetric matrices, particularly if there are multiple eigen-values—remember that these may now be complex numbers. With A $p \times p$, we now have to define two sets of eigenvectors, $\boldsymbol{x}_i$ such that $A\boldsymbol{x}_i = \lambda_i\boldsymbol{x}_i$ and $\boldsymbol{y}_j$ such that $A^{\mathsf{T}}\boldsymbol{y}_j = \lambda_j\boldsymbol{y}_j$. Things are not too bad when all the eigenvalues are distinct. We can then show that the $\boldsymbol{x}$'s are all linearly independent and so are the $\boldsymbol{y}$'s, and also that $\boldsymbol{x}_i^{\mathsf{T}}\boldsymbol{y}_j = 0$ if $i \neq j$, while $\boldsymbol{x}_i^{\mathsf{T}}\boldsymbol{y}_i \neq 0$ and may be standardized to equal 1. If we make the $\boldsymbol{x}$'s into the columns of a matrix X, we have $AX = X\boldsymbol{\lambda}^{\mathrm{d}}$ with $\boldsymbol{\lambda}$ the vector of eigenvalues, and since the $\boldsymbol{x}$'s are independent, then X is nonsingular and $X^{-1}AX$ is diagonal. If the $\boldsymbol{y}$'s are made into the columns of Y in the same way, we have $Y^{\mathsf{T}}A(Y^{\mathsf{T}})^{-1} = \boldsymbol{\lambda}^{\mathrm{d}}$. But we also have the simultaneous transformation $Y^{\mathsf{T}}AX = \boldsymbol{\lambda}^{\mathrm{d}}$ with $Y^{\mathsf{T}}X = I$. Since $X^{-1} = Y^{\mathsf{T}}$ we get the spectral decomposition

$$\begin{aligned} A &= X\boldsymbol{\lambda}^{\mathrm{d}}Y^{\mathsf{T}} \\ &= \lambda_1\boldsymbol{x}_1\boldsymbol{y}_1^{\mathsf{T}} + \lambda_2\boldsymbol{x}_2\boldsymbol{y}_2^{\mathsf{T}} + \cdots + \lambda_p\boldsymbol{x}_p\boldsymbol{y}_p^{\mathsf{T}}. \end{aligned}$$

7.9 Sequences of matrix transforms

Suppose we take a vector $\boldsymbol{z}_0$ and transform it to a new vector $\boldsymbol{z}_1 = A\boldsymbol{z}_0$ where A is $p \times p$ but may be nonsymmetric. We can repeat the procedure as often as we like, getting in general $\boldsymbol{z}_{\mathrm{r}} = A^{\mathrm{r}}\boldsymbol{z}_0$. Suppose that A has unequal eigenvalues λ_i. Then if $\boldsymbol{x}_i$ and $\boldsymbol{y}_i$ are the eigenvectors of A we know that

$$A^{\mathrm{r}} = \lambda_1^{\mathrm{r}}\boldsymbol{x}_1\boldsymbol{y}_1^{\mathsf{T}} + \lambda_2^{\mathrm{r}}\boldsymbol{x}_2\boldsymbol{y}_2^{\mathsf{T}} + \cdots + \lambda_p^{\mathrm{r}}\boldsymbol{x}_p\boldsymbol{y}_p^{\mathsf{T}}$$

and that, since the vectors $\boldsymbol{x}_1, \boldsymbol{x}_2, \ldots, \boldsymbol{x}_p$ span the whole space, we can write $\boldsymbol{z}_0$ as a linear combination of them,

$$\boldsymbol{z}_0 = c_1\boldsymbol{x}_1 + c_2\boldsymbol{x}_2 + \cdots + c_p\boldsymbol{x}_p, \quad \text{say.}$$

Then most of the terms in the product $A^{\mathrm{r}}\boldsymbol{z}_0$ are equal to zero, and we have

$$\boldsymbol{z}_{\mathrm{r}} = c_1\lambda_1^{\mathrm{r}}\boldsymbol{x}_1 + c_2\lambda_2^{\mathrm{r}}\boldsymbol{x}_2 + \cdots + c_p\lambda_p^{\mathrm{r}}\boldsymbol{x}_p$$

Suppose that A has an eigenvalue λ_1 which is greater in modulus than any of the others; we say that such an eigenvalue is *dominant*. Then as r gets larger, the ratio of λ_j^{r} to λ_1^{r} gets smaller and, unless c_1 is exactly equal to zero, z_r will ultimately come to lie along the direction defined by the corresponding eigenvector $\boldsymbol{x}_1$ and to be simply multiplied by λ_1 at each iteration. How long it takes for the effects of the other eigenvalues to die away depends of course on how much smaller they are than λ_1 in modulus.

This and similar results are of interest in the study of Markov chains. Suppose that an individual can be in one of m different states and that at each step of a process the proportion of all the individuals which moves from state i to state j is p_{ij}. These *transition proportions* make up a square matrix P. The group of individuals can be described by the proportions $\pi_{i,t}$ in each of the states at step t of the process, and these in turn make up a vector π_t. Then the vector of proportions at step $t + 1$ is

$$\boldsymbol{\pi}_{t+1} = P\boldsymbol{\pi}_t$$

and in general the proportions at step $t + k$ are

$$\boldsymbol{\pi}_{t+k} = P^k\boldsymbol{\pi}_t$$

If P has a dominant eigenvalue, the long-term state of the group will be described by the corresponding eigenvector, no matter what the initial state π_t.

7.10 Rectangular matrices

Suppose that A is an $m \times n$ matrix, with $m > n$ for convenience in notation, and let X be the orthogonal $n \times n$ matrix containing the standardized eigenvectors of $A^{\mathrm{T}}A$. Then $X^{\mathrm{T}}A^{\mathrm{T}}AX = \boldsymbol{\lambda}^{\mathrm{d}}$, the diagonal matrix containing the eigenvalues. This means that the columns of the $m \times n$ matrix AX are a set of orthogonal vectors. These vectors generate a linear space and geometrically it is clear that we can use them as orthogonal axes in this space. Algebraically, this means that we can find an $m \times m$ orthogonal matrix Y^{T} such that $Y^{\mathrm{T}}AX = M$, where M (possibly after a permutation of the rows) can be written in partitioned form as $[\boldsymbol{\mu}^{\mathrm{d}}, 0]^{\mathrm{T}}$, an $n \times n$ diagonal submatrix $\boldsymbol{\mu}^{\mathrm{d}}$ with one column for each axis alongside a $(m - n) \times n$ submatrix of zeros. It is not difficult to identify Y and M. We have

$$Y^{\mathrm{T}}AX \cdot X^{\mathrm{T}}A^{\mathrm{T}}Y = Y^{\mathrm{T}}AA^{\mathrm{T}}Y = MM^{\mathrm{T}}$$

so that Y contains the eigenvectors of AA^{T} and MM^{T} contains its eigenvalues. At least $m - n$ of these are zero; the remainder are equal to the eigenvalues of $A^{\mathrm{T}}A$ (see Example 7.4) and the elements of $\boldsymbol{\mu}$ are just the square roots of these nonnegative quantities. The expression of A in the form YMX^{T} is called its *singular-value decomposition*, and the elements of $\boldsymbol{\mu}$ are called its singular values.

Singular values can be used to advantage in several statistical problems. For example, let $\boldsymbol{\nu}$ be the vector whose ith element is $1/\mu_i$ if $\boldsymbol{\mu}_1 \neq 0$ and zero if $\mu_i = 0$. Then the matrix $N = [\boldsymbol{\mu}^{\mathrm{d}}, 0]$ is a g-inverse of M and it is easy to show that XNY^{T}, an $n \times m$ matrix, is a g-inverse of A. Now let A be a data matrix of x-variables and $\boldsymbol{h}$ the corresponding vector of y-values, and suppose we want the value of $\boldsymbol{b}$ which minimize $(\boldsymbol{h} - A\boldsymbol{b})^{\mathrm{T}}(\boldsymbol{h} - A\boldsymbol{b})$, the standard problem of least squares. A solution is provided by the vector $\boldsymbol{b} = A^{-}\boldsymbol{h}$. In calculating this, we do not need to form the usual matrix of sums of squares and products, and this renders the problem better behaved from the numerical-analysis point of view.

As another application of the singular-value decomposition, take two data matrices X $(m \times n)$ and Y $(m \times p)$ and let us seek for the linear combinations of the x and y variables which have maximum correlation. Orthogonalize the columns of X and Y, writing $X = QR$, $Y = PS$ with Q and P orthogonal—the Gram–Schmidt procedure can be used for this purpose. Then the maximal correlation is the largest singular value of the matrix $Q^{\mathrm{T}}P$ and the coefficients of the linear combinations are the elements of the corresponding vectors. Statistically, this is the problem of *canonical correlations.* The complete set of pairs of linear combinations defined by the vectors are uncorrelated both between and within sets of variables, and have maximal correlations subject to this condition. Canonical analysis of groups of multivariate observations as discussed in Section 7.7 is the special case of this problem in which the y-variables are dummies defining the groups.

An important property of the singular-value decomposition is given by a result due to Eckart and Young. Let A be an $n \times m$ matrix with singular-value decomposition $A = YMX^{\mathrm{T}}$, and let M_k be obtained from M by replacing all but the k largest singular values by zero. Then $B = YM_kX^{\mathrm{T}}$ is the matrix of rank k or less which minimizes $\sum_{i,j}(a_{ij} - b_{ij})^2$; it may be called the *rank-k least-squares approximation* to A.

The sum of squares of differences can be written as

$$\begin{aligned} &\operatorname{tr}[(A - B)(A - B)^{\mathrm{T}}] \\ &= \operatorname{tr}[YY^{\mathrm{T}}(A - B)XX^{\mathrm{T}}(A - B)^{\mathrm{T}}] \\ &= \operatorname{tr}[(M - M')(M - M')^{\mathrm{T}}] \end{aligned}$$

where $M' = Y^{\mathrm{T}}BX$. This is equal to

$$\sum_i(\mu_i - m'_{ii})^2 + \sum_{i \neq j}(m'_{ij})^2$$

where the second sum is over the nondiagonal elements of M'. B and M' have the same rank, and it follows that for a minimum of the sum of squares

(1) all of the off-diagonal elements of M' must be zero;
(2) all but k of the diagonal elements of M' must be zero;
(3) the remaining k diagonal elements of M' must equal the k largest values of μ_i. Thus M' is equal to M_k as defined above.

M. Greenacre (*Theory and applications of correspondence analysis*, Academic Press 1983) shows how this analysis can be generalized to include orthogonality with respect to more general metrics (as in Section 7.7), and derives most of the standard methods of multivariate analysis as special cases.

Examples 7

7.1 The characteristic equation of $A = \begin{bmatrix} 4 & 8 \\ 8 & 25 \end{bmatrix}$ is

$$\lambda^2 - 29\lambda + 36 = 0.$$

Verify that

$$A^2 - 29A + 36I = 0$$

This is an example of the extraordinary *Cayley–Hamilton* theorem, which states that a matrix satisfies its own characteristic equation.

7.2 (a) Show that the eigenvalues of a projector (a symmetric idempotent matrix) are all equal to 1 or 0, and that the rank of a projector is equal to its trace. If P is a $p \times p$ projector, show that $\mathrm{r}(P) + \mathrm{r}(I - P) = p$.

(b) Let $P = \begin{bmatrix} 575 & 270 & 70 & 785 \\ 270 & 540 & 756 & 162 \\ 70 & 756 & 1274 & -266 \\ 785 & 162 & -266 & 1175 \end{bmatrix} /1782.$

Use the square-root method to show that P has nullity 2, and so has two zero eigenvalues. Show that the first column of P is an eigenvector and find its eigenvalue. Hence find the value of the remaining eigenvalue. What does all this tell you about the matrix P?

7.3 If A is singular, verify that a g-inverse can be obtained from its spectral decomposition as described in Section 7.4.

7.4 (a) If X is $n \times p$, show that X, $X^{\mathrm{T}}X$, and XX^{T} all have the same rank. [If X has rank $r<p$, there is a $p \times (p-r)$ matrix C of rank $(p-r)$ such that $XC = O$. Then $X^{\mathrm{T}}XC = O$, so that $\mathrm{r}(X^{\mathrm{T}}X) \leqslant r$. Conversely, $X^{\mathrm{T}}X$ is symmetric, so that if $X^{\mathrm{T}}XC = O$, then $C^{\mathrm{T}}X^{\mathrm{T}}X = O$ and $C^{\mathrm{T}}X^{\mathrm{T}} \cdot XC = O$. It follows that $XC = O$ (why?), so that $\mathrm{r}(X) \leqslant \mathrm{r}(X^{\mathrm{T}}X)$ and the result we want follows. We have actually shown that the columns of X and of $X^{\mathrm{T}}X$ are subject to the same linear constraints. Exactly the same argument holds for X and XX^{T}.]

(b) Show that a nonzero eigenvalue of $X^{\mathsf{T}}X$ is also an eigenvalue of XX^{T}.

[If $X^{\mathsf{T}}X\mathbf{y} = \lambda\mathbf{y}$, then $XX^{\mathsf{T}}(X\mathbf{y}) = \lambda(X\mathbf{y})$. Because $\lambda \neq 0$ we cannot have $X\mathbf{y} = \mathbf{0}$; so λ is an eigenvalue of XX^{T} with $X\mathbf{y}$ eigen to it.]

7.5 Figure 1.1(b) contains the corrected sums-of-square-and-products matrix A of the data matrix in Fig. 1.1(a). Form the triangular square-root matrix U. Compare your result with example 6.5.

[To show how the inverse of the triangular matrix can be calculated, we simultaneously operate on a unit matrix, as follows:

$$\overset{A}{\begin{bmatrix} 208.74 & 27.14 & -24.59 \\ & 386.42 & 100.32 \\ & & 32.94 \end{bmatrix}} \quad \overset{I}{\begin{bmatrix} 1 & 0 & 0 \\ 0 & 1 & 0 \\ 0 & 0 & 1 \end{bmatrix}}.$$

Following the formulae in Section 7.1 we get

$$\overset{U}{\begin{bmatrix} 14.45 & 1.88 & -1.70 \\ 0 & 19.57 & 5.29 \\ 0 & 0 & 1.44 \end{bmatrix}} \quad \overset{(U^{\mathsf{T}})^{-1}}{\begin{bmatrix} 0.0692 & 0 & 0 \\ -0.0066 & 0.0511 & 0 \\ 0.1061 & -0.1881 & 0.6957 \end{bmatrix}}$$

(one or two more decimal places would be preferable).

The inverse of A can now be found from the sums of squares and products of the columns of $(U^{\mathsf{T}})^{-1}$:

$$\begin{bmatrix} 0.0161 & -0.0203 & 0.0738 \\ \cdot & 0.0380 & -0.1309 \\ & & 0.4840 \end{bmatrix}.$$

This is not the most streamlined approach—it is not too difficult to calculate A^{-1} directly from U without forming U^{-1}—but it is very easy to remember and perform.]

7.6 Consider a number of female organisms classified into several age groups and suppose that the initial number in group i is n_i^0, for $i = 1, 2, \ldots, k$. Suppose that in one unit of time a proportion p_i survives from age group i to age group $i + 1$ (with $p_k = 0$), and that the organisms in group i have f_i female offspring which enter group 1. Find the numbers after one unit of time.

[The numbers in time interval 1 are given by

$$\begin{aligned} n_1^1 &= f_1 n_1^0 + f_2 n_2^0 + \cdots + f_k n_k^0, \\ n_2^1 &= p_1 n_1^0, \\ n_3^1 &= p_2 n_2^0, \\ &\qquad \cdots \\ n_k^1 &= p_{k-1} n_{k-1}^0. \end{aligned}$$

In matrix notation

$$\boldsymbol{n}^1 = A\boldsymbol{n}^0$$

where A is the $k \times k$ matrix

$$\begin{bmatrix} f_1 & f_2 & \cdots & f_{k-1} & f_k \\ p_1 & 0 & \cdots & 0 & 0 \\ 0 & p_2 & & \cdots & 0 \\ \vdots & & \cdots & & \vdots \\ 0 & \cdots & & p_{k-1} & 0 \end{bmatrix}.$$

If A has a dominant eigenvalue, the relative numbers in the age groups will tend to a constant pattern no matter what the starting point. Demographers refer to this pattern as a *stationary population.* If there is no dominant eigenvalue, quite bizarre behaviour is possible. An example is provided by the 3×3 matrix in example 1.4. The characteristic equation of this matrix is $\lambda^3 - 1 = 0$ so that the three eigenvalues all have modulus equal to 1. Since $A^3 = I$, the age pattern repeats itself every third time interval.]

7.7 If A is symmetric $p \times p$ and B is $p \times q$ with $q < p$, show that

$$(A + BB^{\mathrm{T}})^{-1} = A^{-1} - A^{-1}B(I_q + B^{\mathrm{T}}A^{-1}B)^{-1}B^{\mathrm{T}}A^{-1}.$$

This result is useful in practice if A^{-1} is known or easy to find and q is small. Consider for example a multinomial distribution with $k + 1$ categories, and probabilities

$$[\pi_0 \quad \pi_1 \quad \ldots \quad \pi_k] = [\pi_0 \quad \boldsymbol{\pi}]$$

say. If $\boldsymbol{p}$ is the vector of observed proportions in categories 1 to k in a sample of size n, its variance–covariance matrix is

$$n(\boldsymbol{\pi}^{\mathrm{d}} - \boldsymbol{\pi}\boldsymbol{\pi}^{\mathrm{T}}).$$

The inverse of this is immediately given by

$$(\boldsymbol{\pi}^{-\mathrm{d}} + \mathbf{1}\mathbf{1}^{\mathrm{T}}/\pi_0)/n.$$

8
Matrix calculations

8.1 Multiplication

Matrix arithmetic is one of the best understood branches of numerical analysis. Algorithms for the standard operations exist which are both precise and efficient and these are available as procedures written in high-level languages or even as built-in commands in languages such as APL or some dialects of BASIC. The statistical programmer is entitled to assume the existence of such procedures (a rather minimal set will be found in an appendix to this chapter) and should normally not waste time in reinventing them. For those who like to look inside the black boxes, this chapter gives a brief account of some of the problems and how they have been overcome. It is not intended as a programmer's guide; really good algorithms demand an attention to fine detail which would here be inappropriate.

The simple operations of addition, subtraction, and scalar multiplication of matrices require little comment, being accomplished by straightforward loops. Every element of a matrix product is the scalar product of two vectors, and the operation of taking a scalar product is the fundamental building block of matrix arithmetic. Now the scalar product of two vectors with p elements involves p scalar multiplications and p additions and the resulting rounding errors may become nonnegligible. It is often advisable to calculate the scalar multiplications and additions in double precision and to round the total back to single precision at the end, incurring only a single rounding error. This may sometimes reduce the need to use double-precision arithmetic throughout the calculations, though this will usually be necessary on machines with 32-bit words and crude rounding mechanisms.

8.2 Matrix inversion and simultaneous equations

Most of the efficient methods for solving equations and inverting matrices operate in two stages; first the matrix is reduced to triangular form and then the solution is straightforwardly completed (we assume that we are dealing with square, but possibly singular, matrices). One such method, of wide applicability, is *Gaussian elimination*. Suppose we wish to solve the equations

$$\begin{aligned} a_{11}x_1 + a_{12}x_2 + \cdots + a_{1p}x_p &= h_1, \\ a_{21}x_1 + a_{22}x_2 + \cdots + a_{2p}x_p &= h_2, \\ &\cdots \\ a_{p1}x_1 + a_{p2}x_2 + \cdots + a_{pp}x_p &= h_p. \end{aligned}$$

Choose a nonzero element as *pivot* and permute the rows and columns to bring this to the top left-hand corner—we suppose that this has been done. We can now subtract a_{k1}/a_{11} times the first equation from the kth equation for $k = 2, 3, \ldots, p$. Now all the equations except the first involve only $x_2, x_3, \ldots, x_p$. Next we choose another pivot from the rows and columns excluding the first and repeat the process, noting that this leaves unchanged the zeros in the first column. If the coefficient matrix A is nonsingular, we never run out of nonzero pivots and the outcome is a set of equations which we can write as

$$\begin{aligned} b_{11}x_1 + b_{12}x_2 + \cdots + b_{1p}x_p &= k_1, \\ b_{22}x_2 + \cdots + b_{2p}x_p &= k_2, \\ \cdots, \\ b_{pp}x_p &= k_p, \end{aligned}$$

with all the b_{ii} nonzero. These can now be solved from the bottom upwards without difficulty. For the sake of precision it is important to choose the numerically largest available element for pivot at each stage, ensuring that all the multipliers in the process are less than 1. If we require A^{-1} we may set $\boldsymbol{h}$ equal successively to $[1, 0, \ldots, 0]^T$, $[0, 1, \ldots, 0]^T, \ldots, [0, 0, \ldots, 1]^T$ to obtain the p columns of the inverse as the p sets of solutions.

If A is symmetric and positive definite, a good deal of simplification is possible. It is no longer necessary to search for large pivots; the diagonal elements can be used with no loss of precision. The elimination process can be extended so that zeros are produced in both a row and a column at each step; this is called the *Gauss–Jordan* method. This procedure is closely analogous to matrix inversion by partitioning (Section 3.4) with its statistical parallels. In a multiple regression context it can be used to add new x-variables to a regression equation, and a minor variant allows also for the removal of variables (though the numerical behaviour of this last procedure can be dangerous).

An alternative involves the Cholesky reduction of A to the product U^TU with U upper-triangular (Section 7.1). It is simple to produce $V = U^{-1}$ and hence $A^{-1} = VV^T$. If instead we write $A = U^TDU$ with D diagonal and U

an upper triangle with 1's on the diagonal, the square roots can be avoided with some increase in efficiency.

8.3 The precision of the results

The calculated solution or inverse will of course not coincide with the true solution or inverse owing to the calculations being done with only finite precision. The agreement between the two depends upon the *condition* of the matrix, best measured by the ratio of the modulus of the largest to that of the smallest eigenvalue. If this is very large—if, in this sense, the smallest eigenvalue is very small, so that the matrix is 'nearly singular'—the calculated solution may be very far different from the exact solution, a fact which gave rise to a certain air of defeatism when it was first established. However, the situation can be looked at in another way. A calculated inverse will be the *exact* inverse of a matrix which is in general different from the one we start with. However, if we use good algorithms such as those described above, the differences in corresponding elements can be guaranteed not greater than a small multiple of the smallest amount possible, 1 in the last place carried. In this sense, the calculated solutions are about as good as they could possibly be. The fact is that, with extreme ill-conditioning, the smallest possible change in the matrix to be inverted may produce arbitrarily large changes in the exact inverse. When we recall that the elements of the matrix to be inverted probably differ from their 'true' values (if only by having been converted from decimal to binary) we see that we are in a situation where the data provided simply do not determine the solution required. There is a close analogy with the statistical problems of badly determined regression coefficients which are produced when the coefficient matrix of the normal equations is ill-conditioned.

8.4 Generalized inverses

Both the methods described can be used, in theory very simply, to find a g-inverse of a singular matrix. If A, the matrix to be inverted, is $p \times p$ and of rank r, we find that sooner or later the next pivot is zero. When this occurs, we simply have to create a whole row of zeros in the triangular matrix. We can then go on to complete the solution as before. This produces a g-inverse formed by inverting a nonsingular $r \times r$ submatrix and bordering the inverse with zeros.

In practice, things are by no means so simple, because an actual A will almost certainly not be exactly singular, due to rounding and other errors in its elements (the usual textbook examples, with elements that are all small integers, are highly misleading here). As a result, the pivots that 'should' vanish will actually take small nonzero values; the actual matrix to be inverted, rather than being singular, will be just very ill-conditioned. In practice, a somewhat arbitrary line has to be drawn between the two situations, especially when the errors in the original elements are of unspecifiable magnitudes.

8.5 Iterative methods

Simultaneous equations can be solved by various iterative techniques. These can be very convenient for hand calculation, but are less so for computer work, especially as they do not lend themselves to the inversion of matrices. Suppose we have a set of equations $Ax = h$ and a vector $x^0 = [x_1^0, x_2^0, \ldots, x_p^0]^T$ which is an approximate solution. We can insert $x_2^0, \ldots, x_p^0$ into the first equation and use it to get a new value of x_1, say x_1^1. From the second equation, using $x_1^1, x_3^0, \ldots, x_p^0$ we derive x_2^1; from the third equation using $x_1^1, x_2^1, \ldots, x_p^0$ we derive x_3^1; and so on. The vector x will usually converge quite quickly to the solution of the equations. This is the *Gauss–Seidel* method, also known as *relaxation* (just possibly because Gauss claimed that he could use it—without mechanical aids!—while half asleep or thinking about other things).

Gauss–Seidel iteration can sometimes be applied without explicitly setting up the simultaneous equations. Suppose that in a randomized block experiment with b blocks and t treatments a single yield is missing, and the block, treatment, and grand totals involving the missing value are B, T, and G. Then the expected value of the missing entry based on the parameter estimates derived from the rest of the data is given by $(bB + tT - G)/(b - 1)(t - 1)$. Suppose now that p values are missing. Their expected values are the solutions to p simultaneous equations. These can be solved iteratively by inserting approximations to the expected values and using the formula to improve each one of these in turn. This is precisely equivalent to the Gauss–Seidel method.

Another technique starts as before with an approximate solution x^0 and uses $x_2^0, \ldots, x_p^0$ in the first equation to derive x_1^1, then $x_1^0, x_3^0, \ldots, x_p^0$ in the second equation to derive x_2^1, then $x_1^0, x_2^0, \ldots, x_p^0$ in the third equation to derive x_3^1 and so on. Unlike the previous method, the new approximations are not fed in to the process until one whole cycle of iteration is complete. This method converges more slowly than the Gauss–Seidel method, but it is guaranteed to do so, and it may be simpler to implement. Taking the missing-value problem again, now for a quite general design with a single error term, we can insert initial approximations in the missing cells, analyse the data, and form the residuals. The object of the exercise is to produce values in the missing cells which give rise to zero residuals, and the next set of approximations is found simply by subtracting the calculated residuals from the current set. This is a special case of the so-called EM algorithm (Dempster, A. P., Laird, N., and Rubin, D. B. (1977). *J. Roy. Statist. Soc. B* **39**, 1–38).

8.6 Eigensystems of symmetric matrices

If A is a symmetric $p \times p$ matrix, in order to find its eigenvalues and eigenvectors we need an orthogonal matrix X such that $X^{\mathsf{T}}AX$ is diagonal. Now one of the simplest orthogonal matrices P is the unit matrix with four elements altered to the values

$$p_{ii} = \cos\theta \qquad p_{ij} = \sin\theta,$$

$$p_{ji} = -\sin\theta, \qquad p_{jj} = \cos\theta,$$

(for some angle θ) which corresponds to a rotation of the (i, j) axes through an angle θ (compare Section 6.7). If we put $\tan 2\theta = \frac{1}{2}a_{ij}/(a_{ii} - a_{jj})$ and form $P^{\mathsf{T}}AP$, we find that the (i, j) element of the resulting matrix is zero. We can do this successively to zeroize each of the off-diagonal elements in turn; the previous zeros may disappear during subsequent transforms, but the sum of squares of the off-diagonal elements cannot increase and it can be shown that the process converges and the matrix tends to diagonal form. The diagonal matrix contains the eigenvalues and the product of all the rotation matrices contains the eigenvectors. This is called the *Jacobi* method. It is accurate and reliable and very easy to implement. However, it can be rather slow, especially as it insists on finding all the eigenvalues and eigenvectors, which may not be necessary.

If we rotate so as to set to zero only the elements a_{ij} for which $|i - j| > 1$, we find that we can get rid of all of these in a single sweep, without losing any of the zeros once they are formed; this produces a *tridiagonal* matrix, with zeros everywhere except on the main diagonal and on the diagonals immediately adjacent to it. An alternative to the rotation method is to use Householder matrices of the form $I + 2\boldsymbol{u}\boldsymbol{u}^{\mathsf{T}}$ with $\boldsymbol{u}^{\mathsf{T}}\boldsymbol{u} = 1$, as in Section 6.7. In either case the problem of finding the eigenvalues is considerably simplified. The eigenvalues of the original matrix are unchanged by the orthogonal transformations, and the eigenvectors can easily be recovered.

Two methods are in current use for finding the eigensystem of a symmetric tridiagonal matrix. The first takes the original matrix A as the start of an iterative process. Given A_i at some stage of this process, we express this in the form Q_iR_i with Q_i orthogonal and R_i triangular, and then form $A_{i+1} = R_iQ_i = A_i^{\mathsf{T}}A_iQQ_i$. All the A_i are tridiagonal, which greatly reduces the arithmetic, and they tend to a diagonal matrix containing the eigenvalues. The eigenvectors are obtained as before from the product of the Q_i matrices. The rate of convergence of this method depends on the ratio of the moduli of neighbouring eigenvalues. This is usually much too slow for comfort, but it can be very greatly speeded up by operating on a suitably shifted matrix $A_i - k_iI$, where k_i can be obtained in the course of the iteration. With this variant, only two or three iterations per eigenvalue are usually needed. This *QR method* is faster than the Jacobi method, but it still involves finding the whole of the eigensystem.

Another method may be preferable if only a few of the eigenvalues and vectors are wanted, as is often the case in multivariate problems. From the tridiagonal matrix, it is quite easy to determine how many of the eigenvalues are less than a trial value λ. Individual eigenvalues can then be located by a bisection process, and it is possible to ask for (say) the top six eigenvalues ignoring the rest. Eigenvalues lying close together or coinciding, often a source of trouble, are here a positive advantage.

Given an eigenvalue, it is apparently simple to obtain the corresponding eigenvector by solving the appropriate set of simultaneous equations, which with a tridiagonal matrix will never have more than three terms each. This is actually a disastrously bad technique; the calculated eigenvalue will be only approximately correct and the coefficient matrix of the equations will not be singular but merely exceptionally ill-conditioned. Instead we can proceed as follows.

Let C be the tridiagonal matrix and $\hat{\lambda}$ an approximation to the eigenvalue λ_j, and suppose we solve the equations $(C - \hat{\lambda}I)\boldsymbol{x} = \boldsymbol{h}$ where the vector $\boldsymbol{h}$ is for the time being unspecified. We can express $\boldsymbol{h}$ in terms of the eigenvectors of C in the form $\boldsymbol{h} = \Sigma k_i\boldsymbol{y}_i$. Then the eigenvalues of $(C - \hat{\lambda}I)^{-1}$ are $(\lambda_i - \hat{\lambda})^{-1}$, and so the solution is $\boldsymbol{x} = \Sigma k_i(\lambda_i - \hat{\lambda})^{-1}\boldsymbol{y}_i$. If $\hat{\lambda}$ is much closer to λ_j than to any other eigenvalue then $\lambda_j - \hat{\lambda}$ will be very small and $\boldsymbol{x}$ will be nearly coincident with $\boldsymbol{y}_j$, unless k_j is very small too. We therefore reduce C to triangular form by Gaussian elimination with interchanges. Then we assume that we originally had chosen a vector $\boldsymbol{h}$ which would at this stage of the calculations have produced a 1 in each row, and proceed to complete the solution of the equations. This usually gives a very good approximation to $\boldsymbol{y}_j$; it will almost certainly produce a vector which has a reasonable component of $\boldsymbol{y}_j$ in it, and if desired we can use this as the starting point for a repeat of the process.

The main problem remaining is that of getting a set of orthogonal eigenvectors corresponding to a set of coincident eigenvalues. A completely reliable method will be rather complicated. Quite a good method consists in forcing the eigenvalues to be unequal by very small perturbations affecting only the last few bits of each. This is usually sufficient for the resulting vectors to be at least linearly independent, and they can, if necessary be orthogonalized in a subsequent operation.

Appendix: some matrix algorithms

A. 1 Introduction

Although excellent matrix algorithms written in Fortran and Algol 60 are available through the Numerical Algorithms Group library and elsewhere, they tend to be not wholly suitable for statistical applications—like standard matrix algebra textbooks, they allow for contingencies that do not usually arise in statistical work, while failing to cope with situations (such as 'inverting' singular matrices) which are of practical importance.

The algorithms given here are sufficient to cope with most ordinary statistical needs. They are presented in the Ratfor language, of which full details are available in the admirable *Software Tools* by B. Kernighan and P. J. Plauger (Addison-Wesley, 1976). The main advantage of this language is that it is quite readable—much more so than Fortran itself. A few details may be given here.

Ratfor statements can be grouped by enclosing them in braces (curly brackets). Such grouped statements can follow an if or an else; they may form the body of a do loop (equivalent to a for statement in Basic) which thus does not need a label to mark its termination. Two other forms of loop statement are available. The statement

```
for(statement 1; test; statement 2) {...}
```

initializes a loop by obeying statement 1 and making the test. If the test is satisfied, the grouped statements are executed followed by statement 2, and the test is repeated. The loop continues until the test fails. The statement

```
for(i=1; i<=n; i=i+1){...}
```

is almost the same as a do loop, but notice (a) that i is available and equal to $n + 1$ when the loop is finished; and (b) that the loop may be obeyed no times. The for statement is mainly used here for loops which run backwards, or which terminate at $i = n - 1$ rather than $i = n$.

The other loop construct is the statement

```
repeat{...}until(test)
```

This is self-explanatory; the test is made at the end of the loop.

Ratfor statements are written in free format, separated if necessary by semicolons. The # sign means that the rest of the line is to be taken as a comment. The standard symbols >, < are allowed in logical expressions; == means 'is equal to' (the Fortran .EQ.); & means .AND., | means .OR., ! means .NOT..

Programs for translating Ratfor into Fortran are available on most computers, from CP/M micros to Cray 1's (the language is a standard component of the UNIX operating system). It is not at all difficult in practice to translate by hand into a variety of languages since the logic of a Ratfor program is very simple to understand.

A.2 Matrix conventions

Some conventions are needed for the storage of matrices to be processed by the algorithms. All matrices are stored as 1-dimensional arrays; this sounds perverse when a matrix is an essentially 2-dimensional object, but it turns out that the subroutine calls are thereby considerably simplified. Rectangular matrices (such as data matrices) are assumed to consist of real quantities (in the Fortran sense) of standard precision and are stored by columns. Thus a 3×2 matrix is stored in the sequence

$$a_{11}\, a_{21}\, a_{31}\, a_{12}\, a_{22}\, a_{32}.$$

Symmetric matrices are stored as triangles, so that a 3×3 matrix occupies 6 locations in the sequence

$$s_{11}\, s_{21}\, s_{22}\, s_{31}\, s_{32}\, s_{33}.$$

A triangular matrix, whether upper or lower, is stored in the same way.

The algorithms as presented assume that all symmetric and triangular matrices have double precision elements. Symmetric matrices will often contain sums of squares and products or similar quantities, and single precision on 32-bit machines is usually less than adequate to prevent disastrous cancellation on the left at some stage in the calculations.

Two specific points about the algorithms call for mention. First, they are not protected against misuse. Erroneous usage, such as specifications of negative sizes or attempts to find the square root of a matrix with negative eigenvalues, will lead to erroneous results and usually to fatal errors. Secondly, the algorithms ignore a piece of Fortran pedantry which demands the proper dimensioning of dummy arrays; this again permits simpler calling sequences. All Fortran compilers known to me allow this; worried users may change the dimensions of the dummies to an arbitrary large number if they wish.

The routine names are formal rather than meaningful. This avoids possible clashes with other names in the calling program. There are no routines given for addition, subtraction, and multiplication by a scalar, since these operations are easily accomplished by simple loops.

A.3 The algorithms

A.3.1 Output

Two general-purpose output routines are provided, one for double-precision symmetric matrices, the other for single-precision rectangular matrices. Fortran programmers will see that all output is specified as having five decimal places, and that a page width of 80 characters is sufficient. These details can be changed according to taste.

```
subroutine hmt1(s, n, ichan)
#
double precision s(1)
#
do i = 1, n, 5;{       # i counts blocks of 5 columns
  k1 = (i * (i + 1)) / 2
  k2 = k1
  k3 = min0(i + 4, n)
  l = 0
  write(ichan, 201)(k, k=i, k3)
  do j = i, n;{
    write(ichan, 200)j, (s(k), k = k1, k2)
    k1=k1 + j
    l = min0(4, l + 1)
    k2 = k1 + l}
  }
200 format(1x, i4, ' : ', 5f14.5)
201 format(/'0  ', 8i14/)
return
end
```

The double-precision symmetric n × n matrix s is output on channel ichan.

```
subroutine hmt2(a, m, n, ichan)
#
real a(1)
#
do i = 1, n, 5;{
  k3 = min0(i + 4, n)
  write(ichan,201)(k, k = i, k3)
  do j = 1, m;{
    k1 = (i - 1) * m + j
    k2 = (k3 - 1) * m + j
    write(ichan, 200)j, (a(k), k = k1, k2, m)}
  }
200 format (1x, i4, ' : ', 5f14.5)
201 format (/'0  ', 5i14/)
return
end
```

The rectangular real m × n matrix a is output on channel ichan.

A.3.2 Transposition

```
subroutine hmt3(a, m, n, b)
#
real a(1), b(1)
#
na = n * m
do i = 1, na
  b(i) = a(i)
if (m == 1 | n == 1) return
mn = na - 3
do ix = 1, mn;{
  iy = ix
  repeat{
    ic = iy / n
    ir = iy - n * ic
    iy = m * ir + ic}
  until (iy >= ix)
  w = b(ix + 1); b(ix + 1) = b (iy + 1); b(iy + 1) = w}
return
end
```

The rectangular real m × n matrix a is transposed to form b (rectangular n × m). The ingenious method, due to P. F. Windley (*Computer Journal* **2**, 47, 1959) uses a single working location and permits (*pace* Fortran conventions) b to coincide with a.

A.3.3 Multiplication

A single routine copes with different ways of multiplying rectangular matrices.

```
subroutine hmt4(a, nar, nac, b, nb, c, mode)
#
real a(1), b(1), c(1)
#
m = 1
l20 = 1
if (mode == 1 | mode == 3){
  n1 = nar
  n2 = nac
  n3 = nar
  n5 = 1}
else{
  n1 = nac
  n2 = nar
  n3 = 1
  n5 = nar}
if (mode == 1 | mode == 2){
  n4 = 1
  if (mode == 1) n6 = nac; else n6 = nar}
else{
  n4 = nb
  n6 = 1}
#
do i = 1, nb;{
  l10 = 1
  do j = 1, n1;{
    sum = 0.0
    l1 = l10
    l2 = l20
    do k = 1, n2;{
      sum = sum + a(l1) * b(l2)
      l1 = l1 + n3
      l2 = l2 + n4}
    c(m) = sum
    l10 = l10 + n5
    m = m + 1}
  l20 = l20 + n6}
return
end
```

a is rectangular nar × nac. b is rectangular with one dimension equal to nb. c is rectangular real formed as follows:

mode = 1	b is nac × nb,	c (nar × nb)	$= ab$
2	b is nar × nb,	c (nac × nb)	$= a^Tb$
3	b is nb × nac,	c (nar × nb)	$= ab^T$
4	b is nb × nar,	c (nac × nb)	$= a^Tb^T$

The next routine calculates a symmetric matrix product.

```
subroutine hmt5(b, m, n, s, c)
#
real b(1)
double precision s(1), c(1), sum
#
ii = 0
mi = -m
do i = 1, n;{
  ii=ii + i - 1
  iij = ii
  mi = mi + m
  mj = -m
  do j = 1,i;{
    sum = 0.0d0
    kk = 0
    mik = mi
    mj = mj + m
    iij = iij + 1
    do k = 1, m;{
      kk = kk + k - 1
      mik = mik + 1
      mjl = mj
      kl = kk
      do l = 1, k;{
        kl = kl + 1
        mjl = mjl + 1
        sum = sum + dble(b(mik)) * s(kl) * dble(b(mjl))}
      kl = kk + k
      for(l = k + 1; l <= m; l = l + 1){
        kl = kl + l - 1
        mjl = mjl + 1
        sum = sum + dble(b(mik)) * s(kl) * dble(b(mjl))}
      }
    c(iij) = sum}
  }
return
end
```

b is rectangular real $m \times n$. s is double-precision symmetric $m \times m$. c, double-precision symmetric $n \times n$, is formed as $b^T sb$.

A variant of this allows s to be effectively a diagonal matrix, presented in the form of a vector; this produces weighted sums of squares and products.

```
subroutine hmt5a(b, m, n, w, c)
#
real b(1), w(1)
double precision c(1), sum
#
l = 0
i0 = 0
do i = 1, n;{
  jk = 0
  do j = 1, i;{
    sum = 0.0d0
    ik = i0
    do k = 1, m;{
      ik = ik + 1
      jk = jk + 1
      sum = sum + dble(b(ik)) * dble(w(k)) * dble(b(jk))}
    l = l + 1
    c(l) = sum}
  i0 = i0 + m}
return
end
```

b is real rectangular $m \times n$, w is real $m \times 1$. c, double-precision symmetric $n \times n$, is formed as $b^T w^d b$.

Simpler still is the version taking s to be a unit matrix.

```
subroutine hmt5b(b, m, n, c)
#
real b(1)
double precision c(1), sum
#
l = 0
i0 = 0
do i = 1, n;{
   jk = 0
   do j = 1, i;{
      sum = 0.0d0
      ik = i0
      do k = 1, m;{
         ik = ik + 1
         jk = jk + 1
         sum = sum + dble(b(ik)) * dble(b(jk))}
      l = l + 1
      c(l) = sum}
   i0 = i0 + m}
return
end
```

b is real rectangular $m \times n$. c, double-precision symmetric $n \times n$, is formed as $b^T b$.

The next routines provide multiplication of a rectangle by a triangular matrix and by a symmetric matrix respectively.

```
subroutine hmt6(u, m, a, n, b)
#
real a(1), b(1)
double precision u(1), sum
#
mm = (m * (m + 1)) / 2
ll = m * n
l = ll
for(icol = n; icol > 0; icol = icol - 1){
   k0 = mm
   for(irow = m; irow > 0; irow = irow - 1){
      sum = 0.0d0
      kl = ll
      k = k0
      for(j = m; j >= irow; j = j - 1){
         sum = sum + u(k) * a(kl)
         k = k - j + 1
         kl = kl - 1}
      b(l) = sum
      l = l - 1
      k0 = k0 - 1}
   ll = ll - m}
return
end
```

u is upper-triangular double-precision $m \times m$, a is a real rectangular $m \times n$. b, real rectangular $m \times n$, is formed as ua.

```
subroutine hmt7(s, m, a, n, b)
#
real a(1), b(1)
double precision s(1), sum
#
l = 1
j0 = 1
do j = 1,n;{
  i0 = 1
  do i = 1, m;{
    sum = 0.0d0
    i1 = i0
    j1 = j0
    do k = 1, m;{
      sum = sum + s(i1) * a(j1)
      if(k >= i) i1 = i1 + k; else i1 = i1 + 1
      j1 = j1 + 1}
    b(l) = sum
    l = l + 1
    i0 = i0 + i}
  j0 = j0 + m}
return
end
```

s is double-precision symmetric m × m, a is real rectangular m × n. b, real rectangular m × n, is formed as sa.

A.3.4 Inversion and division

The square-root or Cholesky decomposition $s = u^T u$ with s symmetric positive semidefinite and u upper-triangular (see Section 7.1) is used. A fairly crude test for singularity is incorporated.

```
subroutine hmt8(s, n, u, nullty)
#
double precision s(1), u(1), w, tol
data tol /1.0d-8/
#
nullty = 0
j = 1
k = 0
do icol = 1,n;{
  l = 0
  do irow = 1, icol;{
    k = k + 1
    w = s(k)
    m = j
    do i = 1, irow;{
      l = l + 1
      if(i < irow){
        w = w - u(l) * u(m)
        m = m + 1}
      }
    if(irow < icol){
      if(u(l) == 0.0d0) u(k) = 0.0d0
      else u(k) = w / u(l)}
    }
  if(dabs(w) <= dabs(tol * s(k))){
    u(k) = 0.0d0
    nullty = nullty + 1}
  else u(k) = dsqrt(w)
  j = j + icol}
return
end
```

s is double-precision symmetric n × n. u is upper-triangular double-precision such that $u^T u = s$. The quantity nullty is set to the nullity of s (see Chapter 4). The test for singularity is based on the small quantity tol which is set in a data statement.

The matrix inversion routine calls hmt8 to calculate the square-root matrix. A working array is also needed.

```
subroutine hmt9(a, n, c, w, nullty)
#
double precision a(1), c(1), w(1), x
#
call hmt8(a, n, c, nullty)     #form square root matrix
#
nn = (n * (n + 1)) / 2
ndiag = nn
for (irow = n; irow > 0; irow=irow - 1){
  l = ndiag
  if(c(ndiag) == 0.0d0){
    do i = irow, n;{
      c(l) = 0.0d0
      l = l + i}
    }
  else{
    do i = irow, n;{
      w(i) = c(l)
      l = l + i}
    jcol = nn
    mdiag = nn
    for (icol = n; icol >= irow; icol = icol - 1){
      if(icol == irow) x = 1.0d0 / w(irow); else x = 0.0d0
      l = jcol
      for (k = n; k > irow; k = k - 1){
        x = x - w(k) * c(l)
        if (l <= mdiag) l = l - 1; else l = l - k + 1}
      c(l) = x / w(irow)
      mdiag = mdiag - icol
      jcol = jcol - 1}
    }
  ndiag = ndiag - irow}
return
end
```

s is double-precision symmetric n × n. c, double-precision symmetric n × n, is set to a g-inverse of s and nullty to the nullity of s. w is a double-precision array of length at least n used for work space.

The next routine provides a symmetric matrix quotient.

```
subroutine hmt10(s, t, n, u, v)
#
double precision u(1), v(1), s(1), t(1), tol
data tol /1.0d-8/
#
nnn = (n * (n + 1)) / 2
do i = 1, nnn{
  u(i) = s(i)
  v(i) = t(i)}
ii = 1
nn = nnn - n
do i = 1,n;{
  if(u(ii) < tol * s(ii)){       #singularity
    ji = nn + i
    for(j = n; j > 0; j = j - 1){
```

```
        u(ji) = 0.0d0
        v(ji) = 0.0d0
        if (j > i) ji = ji - j + 1; else j = j - 1}
      }
    else{
      v(ii) = v(ii) / u(ii)
      u(ii) = dsqrt(u(ii))
      uii = u(ii)
      ji = nn + i
      for (j = n; j > i; j = j - 1){
        u(ji) = u(ji) / uii
        v(ji) = v(ji) / uii
        ji = ji - j + 1}
      ji = ji - 1      #skip diagonal element - already done
      for (j = j - 1; j > 0; j = j - 1){
        v(ji) = v(ji) / uii
        ji = ji - 1}
      ji = nn + i
      jk = nn + n
      for (j = n; j > i; j = j - 1){
        ki = ji
        for(k = j; k > i; k = k - 1){
          u(jk) = u(jk) - u(ji) * u ( ki)
          v(jk) = v(jk) - u(ji) * v(ki) - u(ki) * v(ji)
          v(jk) = v(jk) + u(ji) * u(ki) * v(ii)
          jk = jk - 1
          ki = ki - k + 1}
        for (; k > 0; k = k - 1){
          v(jk) = v(jk) - u(ji) * v(ki)
          jk = jk - 1
          ki = ki - 1}
        ji = ji - j + 1}
      }
    ii = ii + i + 1}
  return
  end
```

s, t, u, and v are all double-precision symmetric n $\times$ n. u is calculated such that s = u^Tu and v is set equal to w^Ttw where w is a g-inverse of u. It is assumed that the column space of t is within that of s.

For solving a triangular system of equations we need

```
subroutine hmt11(u, m, a, n, b)
#
real a(1), b(1)
double precision u(1), sum
#
mm = (m * (m + 1)) / 2
11 = m * n
1 = 11
for (icol = n; icol > 0; icol = icol - 1){
  k0 = mm
  for (irow = m; irow > 0; irow = irow - 1){
    sum = a(1)
    k1 = 11
    k = k0
    for (j = m; j > irow; j = j - 1){
      sum = sum - u(k) * b(k1)
      k = k - j + 1
      k1 = k1 - 1}
    if (u(k) == 0.0d0)b(1) = 0.0d0; else b(1) = sum / u(k)
    1 = 1 - 1
    k0 = k0 - 1}
  11 = 11 - m}
return
end
```

u is upper-triangular double-precision m × m, a is real rectangular m × n. b real rectangular m × n is calculated as va where v is a g-inverse of u. It is assumed that the equations are consistent.

A.3.5 Eigensystems

The routine presented uses the Jacobi technique, which is rather slow, but compact and reliable.

```
subroutine hmt12(s, n, root, vec)
#
real root(1), vec(1)
double precision s(1)
double precision w, x, anorm1, anorm2, thresh, an, rho, sint, cost
double precision two
logical ind
data rho /1.0e-5/      #convergence criterion
data two /2.0d0/
#
an = n
k = 1       #set up unit matrix in vec
for (i = 1; i < n; i = i + 1){
  vec(k) = 1.0
  k = k + 1
  do j = 1, n;{
    vec(k) = 0.0
    k = k + 1}
  }
vec(k) = 1.0
w = 0.0      #get norm of off-diagonal elements
k = 1
do i = 2,n;{
  k = k + 1
  for (j = 1; j < i; j = j + 1){
    x = s(k)
    w = w + x * x
    k = k + 1}
  }
anorm1 = dsqrt(two * w)
anorm2 = anorm1 * rho / an
if(anorm1 <= 0.0d0) return      #matrix is already diagonal
thresh = anorm1
repeat{      #outer loop of iteration
  thresh = thresh / an
  repeat{      #inner loop
    k = 0
    ii = 1
    ind = .true.     #ind goes false if a rotation is performed
    do i = 2,n;{
      k = k + 1        #k points to s(i,j)
      ii = ii + i      #ii to s(i,i)
      jj = 0           #jj to s(j,j)
      for (j = 1; j < i; j = j + 1){
        k = k + 1
        jj = jj + j
        if (dabs(s(k)) > thresh){      #rotate
          ind = .false.
          v1 = s(ii)
          v2 = s(k)
          v3 = s(jj)
          v4 = (v1 - v3) * 0.5
          if (v4 == 0.0) v5 = -1.0
          else{
            v5 = v2 / sqrt(v2 * v2 + v4 * v4)
            if (v4 > 0.0) v5 = -v5}
```

```
          snt = v5 / sqrt(2.0 * (1.0 + sqrt(1.0 - v5 * v5)))
          sint = snt; cost = sqrt(1.0 - snt * snt)
          iii = ii - i
          jjj = jj - j
          ki = (i - 1) * n
          kj = (j - 1) * n
          do l = 1, n;{
            if (l > i) iii = iii + l - 1; else iii = iii + 1
            if (l > j) jjj = jjj + l - 1; else jjj = jjj + 1
            if (l != i & l != j){      #rotate
              w = s(iii); x = s(jjj)
              s(jjj) = w * sint + x * cost
              s(iii) = w * cost - x * sint}
            ki = ki + 1
            kj = kj + 1
            w = vec(ki)
            x = vec(kj)
            vec(kj) = w * sint + x * cost
            vec(ki) = w * cost - x * sint}
          s(ii) = v1 * cost**2 + v3 * sint**2 - two * v2 * sint * cost
          s(jj) = v1 * sint**2 + v3 * cost**2 + two * v2 * sint * cost
          s(k) = (v1 - v3) * sint * cost + v2 * (cost**2 - sint**2)}
        }
      }
    }
    until (ind)       #if no rotations this time round, go on
  }
  until (thresh <= anorm2)
#
#re-order roots and vectors
#
ij = 0
do i = 1, n;{
  ij = ij + i
  root(i) = s(ij)}
do i = 1, n;{
  v = root(i)
  ik = i
  for(j = i + 1; j <=n; j = j + 1){
    if (root(j) > v){
      ik = j
      v = root(j)}
    }     #ik points to the largest remaining root
  if (ik != i){      #else no interchange required
    root(ik) = root (i)
    root(i) = v
    k1 = n * (i - 1)
    k2 = n * (ik - 1)
    do k = 1, n;{
      k1 = k1 + 1
      k2 = k2 + 1
      v = vec(k1); vec(k1) = vec(k2); vec(k2) = v}
    }
  }
return
end
```

s is double-precision symmetric n × n. The eigenvalues of s are placed in root (real n × 1) in descending numerical order and the corresponding eigenvectors in vec (real rectangular n × n).

Further reading

1. General textbooks with a statistical orientation

Graybill, F. A. (1969). *Introduction to matrices with applications in statistics.* Wadsworth, Belmont.

Searle, S. R. (1982). *Matrix algebra useful for statistics.* Wiley, New York.

2. Generalized inverses (Chapter 5)

Rao, C. R. (1962). A note on a generalized inverse of a matrix with applications to problems in mathematical statistics. *J. Roy. Statist. Soc. B* **24**, 152–8.

Pringle, R. M. and Rayner, A. A. (1971). *Generalized inverse matrices with applications to statistics.* Griffin, London.

Rao, C. R. and Mitra, S. K. (1971). *Generalized inverse of matrices and its applications.* Wiley, New York.

3. Linear spaces (Chapter 6)

Kruskal, W. (1975). The geometry of generalized inverses. *J. Roy. Statist. Soc. B* **37**, 272–83.

Halmos, P. R. (1974). *Finite-dimensional vector spaces.* Springer, New York.

4. Numerical methods (Chapter 8)

Wilkinson, J. H. (1965). *The algebraic eigenvalue problem.* Oxford University Press.

Wilkinson, J. H. and Reinsch, C. (1971). *Handbook for automatic computation. Vol. II: Linear algebra.* Springer, Berlin.

Index